Lecture Notes in Computer Science 8163

Commenced Publication in 1973
Founding and Former Series Editors:
Gerhard Goos, Juris Hartmanis, and Jan van Leeuwen

T0240538

Tilmann Rabl   Meikel Poess
Chaitanya Baru   Hans-Arno Jacobsen (Eds.)

Specifying Big Data Benchmarks

First Workshop, WBDB 2012
San Jose, CA, USA, May 8-9, 2012
and Second Workshop, WBDB 2012
Pune, India, December 17-18, 2012
Revised Selected Papers

 Springer

Volume Editors

Tilmann Rabl
University of Toronto, Department of Electric and Computer Science
10 King's College Road, SFB 540, Toronto, ON M5S 3G4, Canada
E-mail: tilmann.rabl@utoronto.ca

Meikel Poess
Oracle Corporation, Server Technologies
500 Oracle Parkway, Redwood Shores, CA 94065, USA
E-mail: meikel.poess@oracle.com

Chaitanya Baru
University of California San Diego, Supercomputer Center
9500 Gilman Drive, La Jolla, CA 92093-0505, USA
E-mail: baru@sdsc.edu

Hans-Arno Jacobsen
University of Toronto, Department of Electric and Computer Science
10 King's College Road, SFB 540, Toronto, ON M5S 3G4, Canada
E-mail: jacobsen@eecg.toronto.edu

ISSN 0302-9743   e-ISSN 1611-3349
ISBN 978-3-642-53973-2   e-ISBN 978-3-642-53974-9
DOI 10.1007/978-3-642-53974-9
Springer Heidelberg New York Dordrecht London

Library of Congress Control Number: 2013956744

CR Subject Classification (1998): H.2, H.3, J.1, C.4

LNCS Sublibrary: SL 3 – Information Systems and Application, incl. Internet/Web
and HCI

Typesetting: Camera-ready by author, data conversion by Scientific Publishing Services, Chennai, India

Printed on acid-free paper

Springer is part of Springer Science+Business Media (www.springer.com)

Preface

Punctuated by the rapid growth in the use of the Internet, both in the number of devices connected globally and the amount of data per device, the world has been in the midst of an extraordinary information explosion over the past decade. As a consequence, society is experiencing a rate of change in dealing with information that is faster than at any other point throughout history. The data originating from social media, enterprise applications, and computer devices in general, commonly referred to as big data, continue to grow exponentially establishing enormous potential for extracting very detailed information. Big data are often differentiated from traditional large databases using the three Vs: volume, variety, and velocity. Some also include a fourth V, namely, value. With new systems, techniques, and algorithms being developed that can deal with these new database characteristics, emerges the need for a standardized methodology for their performance evaluation.

This sparked the idea among a small group of industry and academic experts to establish a series of workshops explicitly with the intention of defining a set of benchmarks for providing objective measures of the effectiveness of hardware and software systems dealing with big data applications. Chaitanya Baru, Tilmann Rabl, Meikel Poess, Milind Bhandarkar, and Raghunath Nambiar formed a Steering Committee to organize these workshops – the Workshop on Big Data Benchmarking series (WBDB) was born. Everybody on the Steering Committee agreed that big data benchmarks, once established in the industry, will facilitate evaluation of alternative solutions and provide for comparisons among different solution approaches. The benchmarks need to characterize the new feature sets, enormous data sizes, large-scale and evolving system configurations, shifting loads, and heterogeneous technologies of big data and cloud platforms. There are new challenges and options in software for big data such as SQL, NoSQL, and the Hadoop software ecosystem; different modalities of big data, including graphs, streams, scientific data, and document collections, etc; new options in hardware including, HDD vs. SSD, different types of HDD, SSD, and main memory, and large-memory systems; and, new platform options that include dedicated commodity clusters and cloud platforms.

WBDB workshops enable invited attendees to extend their view of big data benchmarking as well as communicate their own ideas. This is accomplished through an open forum of discussions on a number of issues related to big data benchmarking – including definitions of big data terms, benchmark processes and auditing. Each attendee was asked to submit an abstract about an interesting topic, related to big data benchmarking and to give a 5-minute "lightening talk" during the first half of the workshop. After that the workshop attendees, who covered the core big data benchmark issues, which were identified in the workshops, were invited to submit a full paper to be included in these proceedings. This turned out to be a great structure for the first two workshops, because it brought a lot of ideas into the open and, since many workshop attendees did not know each other, served as an introduction of the workshop attendees. During social time, individuals were able to follow up on ideas that were sparked by the lightening talks.

The First Workshop on Big Data Benchmarking (WBDB 2012), held during May 8–9, 2012, in San Jose, CA, in the Brocade facilities, was attended by over 60 invitees representing 45 different organizations, including industry and academia. It was funded by the NSF and sponsored by Mellanox, Seagate, Brocade, and Greenplum. The topics discussed at the first workshop can be grouped into four topic areas: (1) Benchmark Context; (2) Benchmark Design Principles for a Big Data Benchmark; (3) Objectives of Big Data Benchmarking; and (4) Big Data Benchmark Design.

As far as benchmark context is concerned, the consensus of the benchmark attendees was that a big data benchmarking activity should begin at the application level, by attempting to characterize the end-to-end needs and requirements of big data analytic pipelines. While isolating individual steps in such pipelines, e.g., sorting, is indeed of interest, it should be done in the context of the broader application scenario. Furthermore, a wide range of data genres should be considered including, for example, structured, semi-structured, and unstructured data; graphs (including different types of graphs that might occur in different types of application domains, e.g., social networking versus biological networks); streams; geospatial data; array-based data; and special data types such as genomic data. The core set of operations need to be identified, modeled, and benchmarked for each genre, while also seeking similarities across genres.

Numerous examples of successful benchmarking efforts can be leveraged, such as those from consortia as the Transaction Processing Council (TPC), Standard Performance Evaluation Corporation (SPEC), industry-driven efforts such as VMMark (VMWare) and Top500, and benchmarks like Terasort and Graph500. With respect to design principles, the workshop discussed whether those from existing TPC benchmarks, many of which experience an impressively long shelf life, can be adopted for a big data benchmark or whether new once should be developed. The conclusion was that some design principles should be adopted but that others such as scalability and elasticity seen in big data application require the development of new design principles. One of the more contentious topics was the questions of whether the goal of a big data benchmark should foster innovation or competition, i.e., whether it should serve as a technical and engineering or a marketing tool, which split the room into academic-focused attendees and those from industry. The goals of a technical benchmarking activity are primarily to test alternative technological solutions to a given problem. Such benchmarks focus more on collecting detailed technical information for use in system optimization, re-engineering, and re-design. A competitive benchmark focuses on comparing performance and price/performance (and, perhaps, other costs, such as start-up costs and total cost of ownership) among competing products, and may require an audit as part of the benchmark process in order to ensure a fair competition.

Following the successful First Workshop on Big Data Benchmarking, the second workshop (WBDB 2012.in) was held in Pune, India, on December 16–17, 2012, where the facilities were provided by Persistent Systems and Infosys. Unlike the first workshop, an open call for papers was published for WBDB 2012.in. This was a good decision, since it attracted several submissions from international researchers. However, the participation was restricted to one person per company or institution. Each participant was requested to give a presentation. The Steering Committee and the Program Committee did a great job in inviting a balanced crowd of industrial and academic participants. About half of the 40 participants were local, while the other half came from all around the world. The two-day workshop was organized in four major blocks. The first day started with three longer presentations that showed matured research and results of collaborations that were seeded in the first workshop. The second half of the first day was used for discussing the BigData Top100 idea, a big data-related analogy of the Top500 list of the world's fastest super computers. The second day began with discussion of big data-related hardware solutions and ended with domain-specific topics in the big data landscape.

The major result from these two workshops is the definition of a big data analytics benchmark, BigBench. It extents the well-known decision support benchmark TPC-DS with semi-structured and unstructured data, very common in big data workloads. The two workshops also seeded the idea of forming a consortium for the BigData Top100 list and a biweekly Big Data Benchmarking Community call was established, where big data researchers and practitioners present novel use-cases, problems, and solutions.

In this book, the most mature and interesting contributions from the First and Second Workshop on Big Data Benchmarking were collected. We divided the contributions into four categories. Five papers cover benchmarking, foundations, and tools: "TPC's Benchmark Development Model: Making the First Industry Standard Benchmark on Big Data a Success" explains the methodology the TPC uses to develop benchmarks"; "Data Management: A Look Back and a Look Ahead" provides an overview of the TPC and why some benchmarks were successful and some failed; "Big Data Generation" covers how large amounts of data for large-scale factor big data benchmarks can be efficiently generated using The Parallel Data Generation Framework (PDGF); "From TPC-C to Big Data Benchmarks: A Functional Workload Model" describes how benchmark-relevant elements from an application domain can be used to define benchmarks; and "The Implications of Diverse Applications and Scalable Data Sets in Benchmarking Big Data Systems" explores the influence of experiment scale on performance.

The second category is about domain specific benchmarking. Six papers cover a broad range of specific big data domains. "Processing Big Events with Showers and Streams" discusses different categories of stream data and their use-cases. The paper "Big Data Provenance: Challenges and Implications for

Benchmarking" reviews big provenance solutions and explores strategies for benchmarking them. In the paper "Benchmarking Spatial Big Data," the domain of spatial data is explored and discussed. Scientific datasets and benchmarking of array databases are presented in the paper "Towards a Systematic Benchmark for Array Database Systems." "Unleashing Semantics of Research Data" presents challenges in retrieving big semantic data from research documents. This part of the book concludes with a discussion of graph data and its generation in the paper "Generating Large-Scale Heterogeneous Graphs for Benchmarking."

The third category covers hardware-specific approaches to measuring big data aspects. The paper "A Micro-Benchmark Suite for Evaluating HDFS Operations on Modern Clusters" presents storage benchmarks on HDFS and in "Assessing the Performance Impact of High-Speed Interconnects on MapReduce" different network interconnects are evaluated.

The last category presents a full end-to-end big data benchmark. "BigBench Specification V0.1" contains a detailed description of the big data analytics benchmark BigBench including the full set of queries and the data model with scripts to run the benchmark.

The 14 papers in this book were selected out of a total of 60 presentations at WBDB 2012 and WBDB 2012.in. All papers were reviewed in two rounds. We would like to thank all authors and presenters for making both workshops successful. We thank the reviewers for their commitment and our sponsors for helping to keep both workshops free of charge.

October 2013

Tilmann Rabl
Meikel Poess
Chaitan Baru
Hans-Arno Jacobsen

WBDB 2012 Organization

General Chairs

Chaitanya Baru	San Diego Supercomputer Center
Milind Bhandarkar	Pivotal
Raghunath Nambiar	Cisco
Meikel Poess	Oracle
Tilmann Rabl	University of Toronto

Program Committee

Roger Barga	Microsoft
Dhruba Borthakur	Facebook
Goetz Graefe	HP Labs
John Galloway	Salesforce
Armanath Gupta	San Diego Supercomputer Center
Ron Hawkins	San Diego Supercomputer Center
Tony Hu	Drexel University
Iannis Katsis	UCSD
Tim Kraska	Brown University
Hans-Arno Jacobsen	Middleware Systems Research Group
Stefan Manegold	CWI
Ken Osterberg	Seagate
D.K. Panda	Ohio State University
Scott Pearson	Brocade
Beth Plale	Indiana University
Lavanya Ramakrishann	LLNL
Mohammad Sadoghi	IBM
Chandrashekhar Sahasrabudhe	Persistent Systems
Gilad Shainer	Mellanox
S. Sudarshan	IIT Bombay
Florian Stegmaier	University of Passau
Gopal Tadepalli	Anna University

WBDB 2012 Sponsors

WBDB 2012 Sponsors

NSF
Mellanox
Seagate
Brocade
Greenplum

WBDB 2012.in Sponsors

Computer Society of India
Persistent
NSF
Mellanox
Seagate
Brocade
Greenplum
Infosys

Table of Contents

Benchmarking, Foundations and Tools

Domain Specific Benchmarking

Benchmarking Hardware

End-to-End Big Data Benchmarks

TPC's Benchmark Development Model: Making the First Industry Standard Benchmark on Big Data a Success

Meikel Poess

Oracle Corporation, 500 Oracle Parkway,
Redwood Shores, CA 94065, USA
Meikel.Poess@oracle.com

Abstract. There are many questions to answer and hurdles to overcome before an idea for a benchmark becomes an industry standard. After all technical challenges are solved and a prototype benchmark is created, the question arises on how to turn the prototype into an industry standard benchmark that has broad acceptance in the industry, is credible and sustainable over an extended period of time. The Transaction Processing Performance Council is one of the most recognized industry standard consortia for developing and maintaining industry standard benchmarks. Its philosophy and strict rules have assured acceptance, credibility and sustainability of its benchmarks for the last two decades. In this paper the author shows how the TPC model for developing and maintaining benchmarks can be applied to creating the first industry standard benchmark on Big Data.

Keywords: Benchmark Development, Big Data, Database Systems Standard.

1 Introduction

As with the development of any other software product, the process of turning a benchmark prototype into a product is not trivial. Prototypes are built in order to minimize the risks involved in software development. While it is important to develop a prototype to prove general feasibility of the problem solution, a prototype is often far from being a product because many steps are foregone, such as error handling and logging, scalability testing, input validation, maintenance planning, upgrade planning and documentation. However, for the development of an industry standard benchmark having a prototype is especially important, as it allows those who will eventually support the benchmark to verify whether the benchmark tests their product, i.e. hardware and software. This will greatly increase acceptance of the benchmark. Even after a benchmark is introduced to the market place it needs to be maintained, i.e. adapted to new trends in hardware and software.

T. Rabl et al. (Eds.): WBDB 2012, LNCS 8163, pp. 1–10, 2014.
© Springer-Verlag Berlin Heidelberg 2014

The three most recognized industry standard consortia, namely the Standard Performance Evaluation Corporation (SPEC), the Transaction Processing Performance Council (TPC) and the Storage Performance Council (SPC) have developed their own ways to organize benchmark development, to deal with benchmark evolution, i.e. versioning and to publish benchmark results in a way to assure the above key characteristics of a successful benchmark. The TPC, unlike any other consortia has managed to continue benchmarks over decades while keeping benchmarks comparable. This has given companies the ability to compare benchmark results over a very long time period and across many products. In [2] Karl Huppler, long term chair of the Transaction Processing Performance council has defined the following five key aspects that all successful benchmarks have in common:

- Relevant – A reader of the result believes the benchmark reflects something important
- Repeatable – There is confidence that the benchmark can be run a second time with the same result
- Fair – All systems and/or software being compared can participate equally
- Verifiable – There is confidence that the documented result is real
- Economical – The test sponsors can afford to run the benchmark

In this paper, the author introduces the TPC's concept of organizing its benchmark development and motivates why the TPC is a good candidate for developing and hosting the first industry standard benchmark on Big Data.

The remainder of this paper is organized as follows. Section 2 summarizes the contributions the TPC has made over the last 24 years. Section 3 gives an overview on how the TPC is organized and how consensus is made. It is followed by Section 4, which provides an in-depth analysis of how the TPC is organized, how benchmarks are developed and maintained. Section 5 motivates why the TPC is most qualified to develop and host the first industry standard on Big Data. Section 6 concludes this paper.

2   Historical Overview of TPC Benchmarks

For over 20 years the Transaction Processing Performance Council (TPC) has been very successful in disseminating objective and verifiable performance data for transaction processing systems in general and database management systems (DBMS) in various domains, i.e., Online Transaction Processing (OLTP), Decision Support (DS) and Web Application (APP). See also the historic overview of the TPC by Kim Shanley [5]. The TPC developed the four OLTP benchmark specifications, TPC-A, TPC-B, TPC-C [6] and TPC-E, which to date produced over 1000 benchmark publications. The TPC also developed four decision support benchmark specifications, TPC-D,

TPC-H, TPC-R and TPC-DS, which produced to date over 160 benchmark results and two web benchmark specifications, TPC-W and TPC-App, which produced a total of four results. Recently the TPC added a virtualization benchmark based on existing OLTP and Decision Support benchmarks, TPC-VMS. In addition to these domain specific benchmarks, the TPC also developed pricing and energy specifications that are applicable to all existing benchmarks.

The development time for industry standard benchmarks in the TPC is steadily rising due to increased complexities of benchmarks. TPC's first OLTP benchmark specification, TPC-A, was published in November 1989. Built upon Jim Gray's DebitCredit benchmark TPC-A for the first time formalized the rules, which all vendors had to obey in order to publish a benchmark result. About one year later, TPC-B was born. TPC-B was a modification of TPC-A, using the same transaction type (banking transaction) but eliminating the network and user interaction components of the TPC-A workload. The result was a batch transaction processing benchmark. Both TPC-A and TPC-B counted about 40 pages and used a single, simple, update-intensive transaction to emulate update-intensive database environments. Their transactions access schemas with four tables, all using 1:n relationships. Two years later in June 1992, TPC's third OLTP benchmark specification, TPC-C, was approved after about four years of development. Compared to previous OLTP benchmarks, the TPC-C benchmark is more complex because of its multiple transaction types, more complex database schema with 9 tables, 92 columns and 8 primary and 9 foreign keys and more complex overall execution rules. In the first 15 months after approval TPC-C underwent three major revisions (Version 2 and 3). All three versions included major new parts and concepts but were labeled comparable. After a failed Version 4 attempt, in October 2000 Version 5 of TPC-C was approved. This version was non-comparable to previous versions. TPC-C counts 132 pages. In 2006 TPC-E, TPC's latest OLTP benchmark was approved after 6 years of development. TPC-E further increases complexity, e.g. TPC-E defines 33 tables, 188 columns, 33 primary and 50 foreign keys. It counts 286 pages.

A similar pattern can be found with TPC's decision support benchmarks. TPC-D, TPC's first decision support benchmark, was approved in May 1995 after four years of development. TPC-D underwent one major backward compatible revision in 1998 before it was replaced by TPC-H and TPC-R in 1999. The replacement benchmarks, TPC-H and TPC-R were based on TPC-D with some minor execution rules and query changes [3]. Development took about one year. Because of lack of market support TPC-R was decommissioned in January, 2005 only a couple results were published. The ideas for TPC's latest decision support benchmark, TPC-DS, go back as early as 2000. TPC-DS was published in the beginning of 2012 [4].

The following chart gives an overview of when the development of TPC benchmarks started (dotted lines) and during which periods they were active (solid lines).

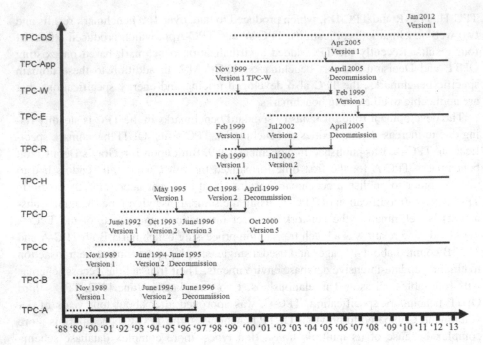

Fig. 1. TPC Benchmark History

3   Organization of the TPC

The TPC is organized hierarchically. The "head" of the TPC organization is the General Council (GC), which takes all decisions during General Council (GC) Meetings, held about every 2 months. During a GC meeting each member company has one vote. In accordance with Robert's Rules of Order a two-thirds vote is required to pass most motions. During these meetings the GC receives feedback from subcommittees in form of subcommittee status reports, which are subsequently distributed via the TPC newsletter to those following the TPC's activity closely.

Directly reporting to the GC are standing subcommittees and technical subcommittees. The standing subcommittees are the Steering Committee (SC), the Technical Advisory Board (TAB) and the Public Relations Committee (PRC). Technical subcommittees are permanent committees that supervise and manage administrative, public relations and documentation issues for the TPC. The Steering Committee (SC) consists of five representatives, elected annually, from member companies. The SC is responsible for overseeing TPC administration and support activities and for providing overall direction and recommendations to the Full Council. Technical Advisory Board (TAB) is tasked with maintaining document and change control over the complex benchmark proposals and methodologies. In addition, the TAB studies issues involving interpretation/compliance of TPC specifications and makes recommendations to the

Council. The Public Relations Committee is tasked with promoting the TPC and establishing the TPC benchmarks as industry standards.

If the GC decides to take on new endeavors, such as developing a new benchmark or defining a term, e.g. processor, it delegates these work items by creating technical subcommittees. Member companies can join and leave subcommittees at any time with approval of the GC. Fig. 2 draws a high level organization chart of the TPC.

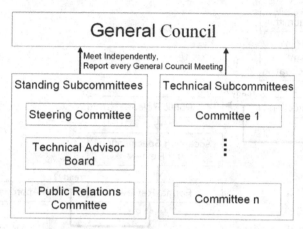

Fig. 2. Hierarchical Structure of the TPC

All major decisions the TPC takes, especially those that affect benchmark specifications, require a super majority to pass. Only procedural decisions, such as voting on committee members require a simple majority. A super majority is defined as two-thirds of the member companies present and voting, excluding abstentions. A simple majority is defined as greater than 50% of member companies present and voting. The super majority rule guarantees strong consensus on important issues among member companies. The hierarchical structure and voting rules set a very high bar for building consensus between otherwise independent entities - mostly corporations. This assures that decisions reached in the TPC are the result of a long consensus process of most member companies, which in itself results in decisions that stand the proof of time.

4   Benchmark Development in the TPC

The TPC has two means to develop benchmark specifications as defined in TPC policies. New benchmark specifications are developed following the *Benchmark Development Cycle*. Revisions of existing benchmark specifications are developed following the *Revisions to a TPC Benchmark Standard*. This section provides a brief review of these two processes and discusses their advantages and disadvantages for the development and maintenance of industry standard benchmarks.

4.1   Benchmark Development Cycle

New benchmark developments must follow the nine-step development process, which is outlined in the figure below. Each box symbolizes one of the nine steps of the development process. The shaded boxes indicate that the general council needs to take action, usually by voting on a motion, the others involve extensive development and testing efforts of technical subcommittee members.

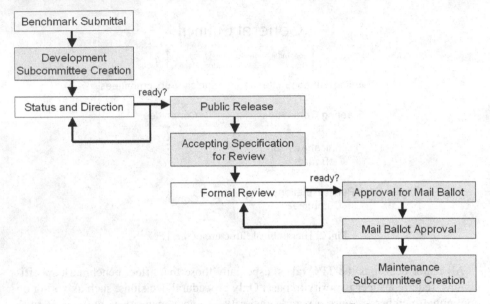

Fig. 3. TPC's current Benchmark Development Cycle

1. Benchmark Submittal:
   Member companies submit an idea for a standard specification in a format similar to TPC Benchmark Standards. If the General Council accepts the proposal a development subcommittee is created to develop the final specification. Depending on the resources that are spent to develop the benchmark idea into a benchmark specification this step can take anywhere between 6 months to 2 years.

2. Creation of a Subcommittee
   This is a formal step taken by the General Council to establish and empower a development subcommittee to develop a formal benchmark specification.

3. Status and Direction
   This step is an iterative process. At each General Council Meeting, which is held approximately every 2 months, the development subcommittee provides a status update on its work, including a working draft of the Specification. During this meeting the Council provides direction and feedback to the subcommittee to further their work.

4. Authorizing Public Release of Draft Specification
Once the General Council is convinced that the specification is almost ready it authorizes the release of a draft Specification to the public. Releasing a specification to the public encourages companies to implement the draft specification, to gather more experimental data, and to speed-up the approval of a specification.

5. Accepting a Standard for Review
When the subcommittee feels that the Specification is of sufficient quality to be considered for formal review and approval, it submits the Specification to the Council for approval to advance into formal review.

6. Formal Review
In this phase, the specification is made available to all TPC members and the public for formal review. All comments and proposed changes generated from the review will be posted in the comment database and considered by the development subcommittee for resolution. This step can take 6 months to 3 years.

7. Approval for Mail Ballot
This is a formal step the General Council takes to approve the updated benchmark specification for mail ballot.

8. Mail Ballot Approval
This is a formal ballot to approve the benchmark specification as a standard. Each member company can submit one vote to either approve, disapprove or abstain the ballot.

9. Creation of a Maintenance Subcommittee
If the mail ballot is approved general council establishes a maintenance subcommittee, which will automatically supersede the development subcommittee.

4.2   Revisions to an Existing Benchmark Specification Standard

The version number of a TPC benchmark specification comprises of three tiers, e.g. the current version of TPC-H is 2.16.0. A revision to an existing benchmark specification can either be a third tier, minor or major revision. Third tier changes clarify confusing or ambiguous areas of the benchmark specification. They do not alter the workload or specification's intent or meaning. Minor revisions entail changes that may alter the workload, intent, or meaning of the benchmark specification. However, the changes are such that benchmark publications that are published under new revision are still comparable to the prior version. Major revision changes alter the workload so significantly or alter the intent of the benchmark specification so drastically such that benchmark publications following the new version are incomparable with older versions.

4.3   General Methodology of TPC Benchmark Specifications

All TPC benchmarks follow a similar methodology and, consequently, follow a similar structure. Each benchmark is technology agnostic. The goal of all TPC benchmarks is to define a set of functional requirements that can be run on any system, regardless of hardware, database management software or operating system. It is the responsibility of those measuring the performance of systems using TPC benchmarks, a.k.a. the test sponsor, to implement the specification and to submit proof that the implementation meets all benchmark requirements, i.e., that the implementation complies with the specification. The proof has to be submitted with every benchmark publication in form of a Full Disclosure Report (FDR). The intent of the full disclosure report is to enable other parties to reproduce the performance measurement. This methodology allows any vendor, using "proprietary" or "open" systems, to implement TPC benchmarks while still guaranteeing end-users that the measurement is comparable. This characteristic differs from benchmarks published from other consortia, most of which provide executable versions of their benchmarks. Those benchmarks are limited to comparing machines that run on a limited number of systems, operating systems and database management systems. In addition, each benchmark results is audited by an independent auditor, who has been certified by the TPC.

TPC benchmarks are modeled after actual production applications and environments rather than being built using synthetic tests. This allows for benchmark analysts to better understand and interpret benchmark results. It also helps to the general reader to relate their workload to the benchmark workload. In addition, testing an actual production application evaluates all key performance factors like user interface, communications, disk I/Os, data storage, and backup/recovery. The challenge in designing such a benchmark, which is supposed to be a standard benchmark representative for a variety of systems and environments, lies in reducing the diversity of operations found in a production application, while retaining its essential performance characteristics, namely, the level of system utilization and the complexity of its operations. A large number of functions have to be performed to manage a production system. Since many of these functions are proportionally small in terms of system resource utilization or in terms of frequency of execution, they are not of primary interest for performance analysis. Although these functions are vital for a production system, within the context of a standard benchmark, they would merely create excessive diversity and expense and are, therefore, omitted. For more detail see [1].

All benchmarks thus far require the database software to provide a minimum set of functionality, e.g. Atomicity, Consistency, Isolation and Durability, commonly referred to as ACID. These system features are tested for each benchmark, not necessarily as part of the performance measurement. TPC-H for instance, defines tests on a smaller database that show ACID compliance.

Before performance numbers can be published, independent auditors must certify their correctness, i.e., the compliance with the specification used. This is done in a certification letter. The audit process may require the auditor to be present during the performance measurement, especially when the systems used have not been benchmarked before. While the general approach to benchmark auditing is identical across TPC benchmarks, each benchmark defines some audit rules that are specific to its application domain. Independent auditors are certified for one or more benchmarks. The certification process for a particular benchmark is conducted by an Auditor Certification Board consisting of TPC members with deep understanding of the benchmark specification. The board reviews the candidate's credentials by conducting an interview including technical questions to verify that the candidate has a solid understanding of the specific benchmark and the technologies and products that can potentially be used in the benchmark implementation and specific questions on the audit requirements of the benchmark. Each TPC benchmark adheres to large extents to the following structure:

Clause	Intention
PREAMBLE	Introduction to the benchmark and high level requirements
DATABASE DESIGN	Requirements and restrictions on how to implement the database schema
WORKLOAD	Detailed definition of the workload
ACID	Atomicity, Consistency, Isolation and Durability requirements
WORKLOAD SCALING	Tools and methodology on how to scale the workload
METRIC/EXECUTION RULES	Detailed description on how to execute the benchmark and how to derive metrics
BENCHMARK DRIVER	Requirements on how to implement the benchmark driver
FULL DISCLOSURE REPORT	Definition on what needs to be disclosed and how to organize the disclosure reports
AUDIT REQUIREMENTS	Minimum requirements for the audit process

5   Big Data Benchmark Development in the TPC

Developing an Industry Big Data Benchmark in the TPC has many advantages as the development process would be based on methodologies refined over 20 years.

The stringent voting rules that are required to pass benchmark changes slow down the development progress, but they also guarantee strong consensus among member companies yielding very long benchmark life spans. Having a benchmark active and results comparable for a long period is attractive because it usually takes the engineering teams a year to familiarize themselves with a new benchmark, i.e., develop a benchmark kit, identify the best hardware and software combination to run the benchmark and potentially develop new hardware and design new algorithms. Hardware and software development cycles are usually measured in years, and since hardware and software vendors are interested in showing continues incremental performance increase from release one release cycle to another it is pertinent that comparable versions of a benchmark are around for reasonable time. Furthermore, TPC benchmarks are complex benchmarks often involving vast amount of engineering resources and capital investments. Hence, vendors are interested in using benchmark results as long as possible for marketing purposes. Long lasting bench-marks also provide a rich data source for performing long term studies as was done in, which compared TPC-C results with Moore's Law [7].

For a benchmark to be successful it needs to generate a strong name recognition. The TPC being around for more than 20 years, who produced thousands of bench-mark results comes with a very strong name recognition.

Benchmark development requires vast amount of engineering resources and capital investments. Without expertise in both benchmark design and knowledge of the un-derlying hardware and software components, which are the focus of the benchmark, a development effort is doomed to fail. The TPC members provide engineers that are experts in benchmark design. They are also very knowledgeable in the products of their companies and, in some cases, also know their competitors product very well.

The TPC grants access to auditing tools, such as the Technical Advisory Board to arbitrate disputes among members. It also provide access to auditors who can be trained to audit any benchmark that.

The TPC also maintains a well known website, which can be used to promote benchmark results. It is frequented by many individuals every day.

References

1. Introduction to the TPC-C benchmark, http://www.tpc.org/tpcc/detail.asp
2. Huppler, K.: Price and the TPC. In: Nambiar, R., Poess, M. (eds.) TPCTC 2010, LNCS, vol. 6417, pp. 73–84. Springer, Heidelberg (2010)
3. Pöss, M., Floyd, C.: New TPC Benchmarks for Decision Support and Web Commerce. SIGMOD Record 29(4), 64–71 (2000)
4. Poess, M., Smith, B., Kollár, L., Larson, P.-Å.: TPC-DS, taking decision support ben-chmarking to the next level. In: SIGMOD Conference 2002, pp. 582–587 (2002)
5. Shanley, K.: Historical overview of the TPC, http://www.tpc.org/information/about/history.asp
6. Current and historic TPC specifications, http://www.tpc.org/tpcc/default.asp
7. Moore, G.E.: Cramming more components onto integrated circuits. Electronics Magazine, 4 (1965) (retrieved November 11, 2006)

Data Management – A Look Back and a Look Ahead

Raghunath Nambiar[1], Ramesh Chitor[2], and Ashok Joshi [3]

[1] Cisco Systems, Inc., 275 East Tasman Drive, San Jose, CA 95134, USA
rnambiar@cisco.com
[2] Cisco Systems, Inc., 260 East Tasman Drive, San Jose, CA 95134, USA
rchitor@cisco.com
[3] Oracle Corporation, 10 Van De Graaff drive, Burlington, MA 01803, USA
ashok.joshi@oracle.com

Abstract. The essence of data management is to store, manage and process data. In 1970, E.F. Codd developed the relational data model and the universal data language "SQL" for data access and management. Over the years, relational data management systems have become an integral part of every organization's data management portfolio. Today, the world is in the midst of an information explosion fueled by worldwide adaption of internet and increase in number of devices connected to the internet. The velocity, volume and velocity of data generated are beyond the capabilities of traditional relational database management systems. This explosive growth has encouraged the birth of new technologies like Hadoop and NoSQL.

This paper gives an overview of the technology trends in data management, some of the emerging technologies and related challenges and opportunities and eminent convergence of platforms for efficiency and effectiveness.

Keywords: Data Management, Big Data, Hadoop, NoSQL.

1 Historical Perspective

The six generations of distinct phases in data management evolving from manual methods, through several stages of automated data management articulated by Jim Gray is depicted in Table 1 [1].

Table 1. Phases in data management

4000 BC	1800	1960	
Paper and Pencil			
	Punch Cards		
		Computers	

Data management has gone through a series of disruptive innovations since the first hierarchical data management systems of the sixties as shown in table 2. These systems were developed when computers became a more cost-effective option for private organizations.

T. Rabl et al. (Eds.): WBDB 2012, LNCS 8163, pp. 11–19, 2014.

Table 2. History of Data Management Technologies

1960s	1970s	1990s	2000-Present
Traditional Files	Hierarchical (CODASYL)	Relational	Big Data

Hierarchical data management systems modeled parent-child relationships between related records. This provided a convenient solution that was rapidly found to have significant shortcomings. Firstly it was an inflexible data model; the database and the application program were very tightly coupled. And once the data hierarchy was embodied in the application, it was not easy to change the data model without changes to the application. Often, it was also necessary to reload the data in a different "format" to reflect the new relationships. Furthermore, the hierarchical data model imposed limits on the kinds of queries that could be executed against the database.

CODASYL systems, named after Conference on Data Systems Languages, is a 1959 consortium that worked on standardizing database interfaces, also better known for developing COBOL programming languages) [2][3] improved upon these limitations by providing the ability to express parent-child relationships as well as other kinds of relationships. These relationships were embodied as "pointers" to the related records in the database. The application had to explicitly update the record in order to establish the relationships with related records. The ability to relate records (using physical pointers) provided more flexibility to the application developer and made it possible to run new kinds of queries without having to completely reorganize the database and/or the application.

Ted Codd at IBM invented the theoretical foundations for relational database systems in the early 70s [4] [5]. RDBMS as they were commonly referred to as, allow the user to establish relationships between two entities using logical relationships (joins). In essence, each row includes information that can logically identify the other rows it is related to. For example, an *Employee* row will contain a *department_id*. The *Department* row also contains a *department_id*. If the value of the *department_id* in an employee row matches the *department_id* in a Department row, then, semantically, that employee is related to (works in) that particular department. However, since the relationships are maintained logically, queries can be expressed declaratively, rather than in a program.

The SQL language (a programming language designed for managing data held in RDBMS) provided powerful tools to quickly and easily manage and query large data sets. This flexibility and ease of use led to tremendous popularity and widespread adoption of relational systems. There was a lot of research in next 15 years on the implementation and performance of RDBMS since the earlier versions did not match the performance of CODASYL systems.

The following years saw huge improvements in performance and functionality of these relational database systems as well as the underlying hardware platforms. Modern database systems are able parallelize queries, use smart heuristics for query optimization, support a wide variety of data formats including text and multimedia and allow the user to express a wide variety of questions simply and easily using SQL.

They also provided a rich set of enterprise-class features such as security, intelligent backups and disaster recovery. Due to these advances, as well as the plethora of RDBMS-based systems available today, the vast majority of enterprise data management and data processing is based on relational database systems.

The 90s saw the emergence of object-oriented databases (or OODBMS, for Object-Oriented Database Management Systems) [6]. They try to make transactions and persistence transparent to the object-oriented programmer. These systems too had their shortcomings, for example, data in a specific OODBMS is typically accessible from a specific programming language using a specific API, which is typically not the case with Relational databases. OODBMS also were less efficient when the data and the relationships were simple. By this time RDBMS were much more standardized and less likely to change.

Perhaps its biggest disadvantage was in an RDBMS modifying the database schema either by creating, updating or deleting tables is typically independent of the actual application. In an OODBMS based application modifying the schema by creating, updating or modifying a persistent class typically means that changes have to be made to the other classes in the application that interact with instances of that class. This typically means that all schema changes in an OODBMS will involve a system wide recompile. It is fair to say that OODBMS did not gain widespread popularity and remain niche solutions to this day.

Historically, CODASYL (to a lesser extent) and relational technologies were disruptive innovations that changed the data management industry. Since the early days, the benefits of using data management systems were obvious. Though earlier systems were hard to use and required custom programming, the cost was justified by the business benefits. Relational technologies and SQL made it radically simpler to develop applications, resulting widespread use of these systems.

2   Big Data

Big Data is another disruptive phenomenon that has emerged in recent years. It is still in the early stages of development, and still suffers many of the same issues and programming difficulties mentioned earlier. However, it is very clear that harnessing its capabilities provides compelling business benefits.

Big Data is a term applied to data sets that are so large that commonly used software tools cannot capture, manage, and process them within a tolerable time in a cost-effective manner. Its data sizes are a constantly moving target, currently ranging from a few dozen terabytes to many petabytes. Examples of Big Data sources vary widely, including web logs, data from radio-frequency ID (RFID) sensor networks, social network information, Internet text and documents, Internet search indexing, and call-detail records. Scientific examples include astronomy and atmospheric science data, as well as genomics, bio-geochemical, biological, and other complex and interdisciplinary scientific research. In addition, military surveillance, medical records, photography archives, video archives, and large-scale ecommerce applications generate huge data volumes. While the types vary widely, the unifying theme is that this data represents potential insight and value for organizations.

Growth of Big Data shows no sign of abating. In a June 2011 report titled Extracting Value from Chaos, IDC estimated that 1.8 zettabytes of data would be created in 2011, growing to 16 zettabytes by 2016 [7]. Now enterprise data is so large that storing and managing it with traditional tools, such as relational database management systems, no longer is economical. As a result, Big Data is increasingly becoming an enterprise-level concern. IDC estimates that individuals create 75 percent of the information in the digital universe, yet enterprises have some liability for 80 percent of this information at some point in its digital life.

3   Information Explosion and Avnet of Internet

The explosive growth of the Internet, wide-area cellular systems and local area wireless networks which promise to make integrated networks a reality, the development of "wearable" computers and the emergence of "pervasive" computing paradigm, are just the beginning of "The Wireless and Mobile Revolution". Today, 30% of world's population has Internet access. The majority of all business interactions are conducted over the Internet. There are 15 billion devices connected to the Internet today; that's more than 2 devices for every human being living on the planet [8]. Wireless broadband is still in its infancy and will continue to grow better and faster in the coming years. This enables people to be more connected and more mobile. If the necessary bandwidth is available, it will always be more efficient to carry around a "battery with a screen" and do all of the data processing on servers in remote data centers, reminiscent of the dumb terminal concept of the yesteryears. It is true that a desktop is more responsive; however, it is important to keep in mind that human users don't need minimal latency; they just need acceptable latency. The realization of wireless connectivity is bringing fundamental changes to telecommunications and computing and profoundly affects the way we network, compute, analyze, communicate, and interact. Interactive communication and information is now available and cheaply accessible to vastly more people than it was ever before.

This emergence of the Internet and mobile applications has led to an explosion of data. Storing all this data and providing simple ways to access it poses huge challenges to the data management tools developers. In the past 15 years, Internet-based companies and Web 2.0 companies such as Yahoo, Google, Twitter and Facebook have had to deal with Big Data out of necessity. The requirements of their business compelled these pioneers to develop innovative and cost-effective solutions for their data management problems. Fortunately, these companies were willing to share the key concepts and lessons learned from these innovations with the developer community. Further, they also established beyond the shadow of any doubt that harnessing Big Data provides significant additional business benefits to the enterprise.

As mentioned earlier, Web 2.0 companies were early innovators of Big Data solutions, resulting in a growing collection of open source technologies that dramatically changed the culture of collaborative software development and the scale and economics of hardware infrastructure. These technologies enable data storage, management and analysis in ways that were not possible with traditional technologies a few years ago. And they are not all are suitable for the enterprise. While these solutions are attractive from the standpoint of the innovation they can bring, many organizations require dependable, supported, and tested enterprise-class solutions for rapid deployment and mission-critical operation.

4   Evolution of Big Data Tools and Technologies

Traditional technologies such as relational database management systems often are unable to handle the volume and velocity of Big Data in a cost-effective manner, resulting in the emergence of two broad categories of technologies – interactive (or real-time) processing and batch processing (analytics).

NoSQL is one such technology that has emerged in the interactive processing space as an increasingly important part of interactive Big Data solutions for applications that need to process large volumes of simple reads and updates against very large datasets. NoSQL is often characterized by what it is not, and definitions vary. It can be *Not Only SQL-based* or simply *Not a SQL-based* database management system. It may not provide full ACID (atomicity, consistency, isolation, durability) guarantees but still has a distributed and fault tolerant architecture.

NoSQL databases form a broad class of non-relational database management systems that are evolving rapidly, and several solutions are emerging with highly variable feature sets and few standards, each suited to address a certain type of interactive Big Data management system (e.g. key-value stores, document stores, columnar stores, etc.). While these technologies are attractive from the standpoint of the innovations they can bring, not all products meet enterprise requirements.

As Big Data processing becomes increasingly important to the success of the business, many organizations require robust, commercially supported solutions for rapid deployments and the ability to integrate such solutions into existing enterprise applications infrastructure.

Map/Reduce computing falls into the category of batch solutions [9]. Rather than focusing on each individual data item, Map/Reduce is designed to derive information by aggregating large amounts of data in multiple ways. For example, a single tweet may not be of much significance; however, if there's a large number of tweets about a topic within a short timeframe, that may provide significant information about an opinion or sentiment by a large population. Map/Reduce technology can be used to quickly derive aggregate information from data. Hadoop popularized by Yahoo, has become synonymous with the Map/Reduce style of data processing. It brings massively parallel computing to commodity servers, resulting in a sizeable decrease in cost per terabyte of storage.

These two categories of Big Data processing are being adopted widely by organizations in order to enter the Big Data era successfully.

5   Advantages of Big Data Technologies

We have already seen the significance of Big Data tools like NoSQL and Map/Reduce. Here are some other noteworthy tools open-sourced under Apache umbrella, for working with Big Data. Big Data technologies have several advantages as listed below.

- **Elastic Scaling:** Historically, database deployments relied on the capability to scale up by deploying bigger servers as database loads increased. Today, technology advancements deliver transparent scaling, enabling organizations to add new servers as business demands dictate. With Big Data solutions, the massive scale-out capabilities of technologies are designed to expand transparently and dynamically, to large numbers of servers in a cost-effective manner, effectively enabling making scaling trivial to design and implement.
- **Economics:** Big Data technologies typically use clusters of inexpensive industry-standard servers to manage rapidly expanding data and transaction volumes, whereas RDBMSs tend to rely on expensive proprietary servers and storage systems. As a result, the cost per gigabyte or transactions per second for can be many times less than the cost for RDBMS, enabling organizations to store and process more data at a much lower price point.
- **Flexible Data Models:** RDBMSs are built on a schema-centric approach in which even minor changes to the data model can result in complex schema. In addition, application changes often necessitate downtime or reduced service levels. Big Data technologies have far more relaxed data model restrictions. Consequently, application and database schema changes do not have to be managed as one complicated change unit.
- **Real-time Customizations:** Consumer companies have long used data to segment and target customers. Big Data creates a whole new playing field by making real-time personalization possible. An oft-quoted example is a retailer being able to track the behavior of individual customers from Internet click streams, update their preferences, and model their likely behavior in real time. They will then be able to recognize when customers are nearing a purchase decision and nudge the transaction to completion by bundling preferred products, usually offered with reward programs.
- **Risk Calculations for Large Portfolios:** The past few years have been anything but smooth sailing for financial services firms that have struggled to effectively manage their portfolios. An industry-wide failure to properly assess the latent risks lurking in thousands of substandard loans led to billions of dollars of losses. For a major company in the financial services market, one of the root causes of its unacceptable risk exposure was simply an inability to efficiently create models and run those models against its growing data volumes. The institution pursued and deployed a new paradigm for its analytical processing: Big Data analytics processing. This reduced the wait time from a week to a few minutes, translating to savings of millions of dollars.
- **Improve Decision Quality:** Big Data when leveraged in a timely manner can provide insights from the vast amounts of data. This includes those

already stored in company databases, from external third-party sources, the Internet, social media and remote sensors. These insights and information can improve decision quality.

6   Big Data in Enterprise

We have already seen the significance of Big Data tools. Organizations globally are beginning to explore how Big Data can be used in order to improve the business. According to one important study, companies taking advantage of the superabundance of data through "data-directed decision-making" enjoy up to 6 percent productivity improvements. As described in another insight from the IBM 2010 Global CFO Study [11], over the next three years, organizations that leverage Big Data will financially outperform their peers by 20 percent or more. The only way to keep up with the expanding data is to think beyond traditional RDBMS tools in a way that most enterprises have never done.

Just as relational database systems were adopted widely and across a wide variety of industries and applications, Big Data brings additional value to a wide variety of organizations and enterprises, both big and small, regardless of the industry they cater to. Through analysis of the large volumes of data there is the potential for making faster advances in many scientific disciplines and improving the profitability and success of many enterprises [12]. Few relevant use cases where Big Data technologies can play a significant role are listed below:

- **Data Storage:** Collect and store unstructured and semi-structured data in a fault-resilient scalable data store that can be organized and sorted for indexing and analysis.
- **Credit Scoring:** Update credit scoring models using cross-functional transaction data and recent outcomes, to respond to changes such as bubble markets collapsing. Sweep recent credit history to build transactional/temporal models.
- **Data Archive:** Medium-term archival of data from EDW/DBMS to increase the length of time that data is retrained or to meet data retention policies/compliance.
- **Integration with Data Warehouse:** Transfer data stored in Hadoop to and from a separate DBMS for advanced analytics.
- **Customer Risk Analytics:** Build a comprehensive data picture of customer-side risk based on activity and behavior across products and accounts.
- **Personalization and Asset Management:** Create and model investor strategy and goals based on market data, individual asset characteristics, and reports fed into online recommendation system.
- **Retailer Compromise:** Prevent or catch frauds resulting from a breach of retailer cards or accounts by monitoring, modeling, and analyzing high volumes of transaction data and extracting features and patterns.

- **Mis-categorized Fraud:** Reduce false positives and prevent mis-categorization of legitimate transactions as fraud, using high volumes of de-normalized data to build good models.
- **Next-Generation Fraud:** Daily cross-sectional analysis of portfolio using transaction similarities to find accounts that are being cultivated for eventual fraud, using common application elements, temporal patterns, vendors and transaction amounts to detect similar accounts pre-bust-out.
- **Social Retention:** Combine transactional, customer contact information and social network data to do attrition modeling to learn social and transaction markers for attrition and retention.
- **Sentiment and Bankruptcy:** Find better indicators to predict bankruptcy among existing customers using sentiment analysis from social networking, responding quickly before the warning horizon.

7   Conclusion and Outlook

Database systems have continued to evolve progressively over the years. But Big Data has dramatically changed how data is envisioned, managed and leveraged, and in enterprise world its potential is just beginning to be understood. This is an exciting time with a large number of tools and technologies being developed for Big Data and large datasets. More and more enterprises are being forced out of their traditional way of working with datasets. Led by Web 2.0 companies, exploitation of large datasets in innovative ways is becoming increasingly common. Interesting trade-offs involving data storage and retrieval, search and heuristic analysis, availability, and data consistency are being made all the time, and, as to be expected, differently in different enterprise domains.

Understandably, with Big Data being the bleeding edge, the tools and technologies available to effectively leverage its potential are still in nascent stage of development. Enterprises with large datasets, can only neglect, the challenges and the opportunities presented by Big Data technologies at their own peril.

References

1. Gray, J.: Data Management: Past, Present, and Future,
   http://ftp.research.microsoft.com/pub/tr/tr-96-18.pdf
2. CODASYL: Conference on Data Systems Languages Records, 1959-1987
3. CODASYL, http://en.wikipedia.org/wiki/CODASYL
4. Codd, E.F.: Relational Completeness of Data Base Sublanguages. Database Systems: 65–98. CiteSeerX: 10.1.1.86.9277
5. Codd, E.F.: The Relational Model for Database Management, version 2nd edn. Addison Wesley Publishing Company, ISBN 0-201-14192-2
6. OODBMS, http://en.wikipedia.org/wiki/OODBMS
7. Extracting Value from Chaos, IDC 2011 (2011),
   http://www.emc.com/collateral/analyst-reports/idc-extracting-value-from-chaos-ar.pdf

8. Nambiar, R.: Information Explosion a Storage Perspective,
   http://gold.cs.pitt.edu/sites/gold.cs.pitt.edu.seedm/slides/
   invited/Nambiar.pdf
9. Map/Reduce, http://research.google.com/archive/mapreduce.html
10. Hadoop, http://hadoop.apache.org/
11. IBM Global CFO Study,
    http://www-935.ibm.com/services/us/cfo/cfostudy2010/
12. Challenges and Opportunities with Big Data, http://www.cra.org/ccc/files/
    docs/init/bigdatawhitepaper.pdf

Big Data Generation

Tilmann Rabl and Hans-Arno Jacobsen

Middleware Systems Research Group
University of Toronto
tilmann.rabl@utoronto.ca, jacobsen@eecg.toronto.edu
http://msrg.org

Abstract. Big data challenges are end-to-end problems. When handling big data it usually has to be preprocessed, moved, loaded, processed, and stored many times. This has led to the creation of big data pipelines. Current benchmarks related to big data only focus on isolated aspects of this pipeline, usually the processing, storage and loading aspects. To this date, there has not been any benchmark presented covering the end-to-end aspect for big data systems.

In this paper, we discuss the necessity of ETL like tasks in big data benchmarking and propose the Parallel Data Generation Framework (PDGF) for its data generation. PDGF is a generic data generator that was implemented at the University of Passau and is currently adopted in TPC benchmarks.

1 Introduction

Many big data challenges begin with extraction, transformation and loading (ETL) processes. Raw data is extracted from source systems, for example, from a web site, click streams (e.g. Netflix, Facebook, Google) or sensors (e.g., energy monitoring, application monitoring, traffic monitoring). The first challenge in extracting data is to keep up with the usually very data high production rate. In the transformation step, the data is filtered and normalized. In the last step, data is finally loaded in a system that will then do the processing. This preprocessing is often time-consuming and hinders an on-line processing of the data. Nevertheless, current big data benchmarks, e.g. GraySort [1], YCSB [2], HiBench [3], BigBench [4], mostly concentrate on a single performance aspect rather than giving a holistic view. They neglect the challenges in the initial ETL processes and data movement. A comprehensive big data benchmark should have an end-to-end semantic considering the complete big data pipeline [5]. An abstract example of a big data pipeline as described in [6] is depicted in Figure 1.

Current big data installations are rarely tightly integrated solutions [7]. Thus, a typical big data pipeline often consists of many separate solutions that cover one or more steps of the pipeline. This creates a dilemma for end-to-end benchmarking. Because many separate systems are involved an individual measure for each part's contribution to the overall performance is necessary for making purchase decisions for an entire big data solution. A typical solution to this dilemma

T. Rabl et al. (Eds.): WBDB 2012, LNCS 8163, pp. 20–27, 2014.

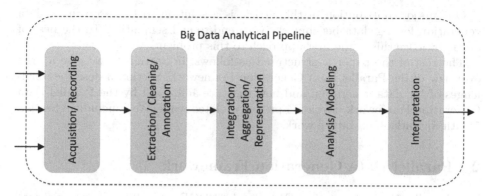

Fig. 1. Abstract Stages of a Big Data Analytics Pipeline

is a component based benchmark. This requires having separate benchmarks for different stages of the big data pipeline. An example is HiBench [3], which includes separate workloads and micro-benchmarks to cover typical Map-Reduce jobs. HiBench, for example, includes workloads for sorting, clustering, and I/O. These micro-benchmarks are run separately and, consequently, inter-stage interactions, i.e., interference and interaction between different stages, as they would appear in real-live systems, are not reflected in the benchmarks.

Considering inter-stage interactions makes the specification of an end-to-end benchmark challenging. This is because it is supposed to be technology agnostic, i.e., it should not enforce a certain implementation of the system under test and also not enforce fixed set of stages. A benchmark should challenge the system as a whole. This creates a dilemma for end-to-end benchmarking of a big data pipeline, because an end-to-end benchmark should not be concerned about the individual steps of the pipeline, which can differ from system to system, but all steps should be stressed during a test. A solution to this dilemma is a benchmark pipeline, where intermediate steps are specified but not enforced and only the initial input and final output are fixed.

For a benchmark to be successful it has to be easy to use. Benchmarks that come with a complete tool chain are used more frequently than benchmarks that consist only of a specification. A recent example is the YCSB, which has gained a lot of attention and a wide acceptance. YCSB is used in many research projects as well as in industry benchmarks (e.g., [8,9]). For a big data benchmark the most important and challenging tool is the data generator. In order to support the various steps of big data processing, it would be beneficial to have a data generator that can generate the data in different phases consistently. This makes a verification of intermediate results as well as isolate single steps of the benchmark procedure possible and thus further increases the benchmarks applicability. In such a data generator the data properties that are processed (such as dependencies and distributions) need to be strictly computable. A data generation tool that follows this approach is the Parallel Data Generation Framework.

Our major contribution in this article is a solution to the problem of data generation for big data benchmarks with end-to-end semantics. To the best of our knowledge this is the first approach to this problem.

The rest of the paper is structured as follows, in Section 2, we give a brief overview of the Parallel Data Generation Framework. Section 3 describes challenges of big data generation and how they are addressed by the Parallel Data Generation Framework. Section 4 presents related work. We conclude in Section 5 with an outlook on future work.

2   Parallel Data Generation Framework

The Parallel Data Generation Framework (PDGF) is a flexible, generic data generator that can be used to generate large amounts of relational data very fast. It was initially developed at the University of Passau and is currently used in the development of an industry standard ETL benchmark (described in [10]). PDGF exploits parallel random number generation for an independent generation of related values. The underlying approach is straight forward; the random number generator is a hash function which can generate any random number in a sequence in O(1) without having to compute other values. Random number generators with this feature are, for example, XORSHIFT generators [11]. With such random number generators every random number can be computed independently. Based on the random number arbitrary values can generated using mapping functions, dictionary lookups and such. Quickly finding the right random number is possible by using a hierarchical seeding strategy (table → column → row).

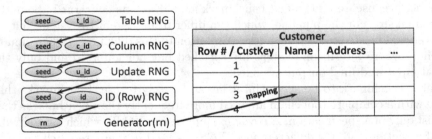

Fig. 2. PDGF's Hierarchical Seeding Strategy

An overview of PDGF's seeding strategy can be seen in Figure 2. The seeding strategy starts by assigning a random number to each table, this number is used as a seed for each column random number generator. PDGF is capable of generating consistent updates, i.e, inserts, deletes, and updates in an abstract time interval (for details refer to [12]). Which and how values are updated is determined by the update random number generator, the resulting seeded row value random number generator is used to deterministically compute the random numbers required for the actual value generation. Having a seeded random number generator for the value generation instead of a single random number or fixed number of values makes it possible to generate values that use a non-deterministic number of random numbers, such as text.

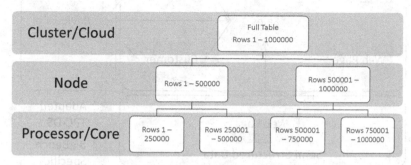

Fig. 3. Parallel Data Generation in PDGF

Not all values should be randomly chosen. An example are references. For tables that contain foreign key constraints, for example, the keys must exist in the referenced tables, which is challenging in the case of non-dense keys or multi-part keys. Using the deterministic approach, existing values can easily and efficiently be recomputed. Furthermore, being able to independently generate all values makes it possible to fully parallelize and distribute the data generation. This especially interesting for big data applications. PDGF comes with an integrated scheduling system that automatically handles multi-core and multi-node parallelism. The working principle is presented in Figure 3. Each table can be split up in equal sized partitions, which can be generated on shared nothing machines. Each partition can further be divided up in multiple subsets, which can be distributed to separate threads or processes.

For further details of this generation approach see [12,13,14,15,16,17].

3   A Big Data Generator

One can build a versatile data generator for big data benchmarking based on PDGF. Although PDGF was built for relational data it features a post-processing module that enables a mapping to other data formats such as XML, RDF, etc. Since all data is deterministically generated and the generation is always repeatable it is possible to compute intermediate and final results of transformations. The underlying relational model makes it possible to generate consistent queries on the data. This makes PDGF an ideal candidate tool for big data benchmarking.

PDGF was recently used for generating the data set for the BigBench big data analytics benchmark [4]. BigBench models a retail business, were articles are sold in stores and over websites. The schema consists of structured, semi-structured, and unstructured data as can be seen in Figure 4. The structured core of the schema is adapted from TPC-DS [18]. The semi- and unstructured parts are

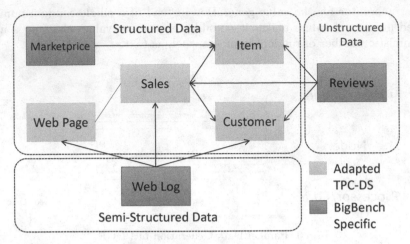

Fig. 4. BigBench Schema

implemented in PDGF. The unstructured part models reviews of products. The semi-structured part models an Apache Web server log. The reviews are used for sentiment analysis, which requires very realistic text in order to get reasonable results. This is achieved by using Markov chains.

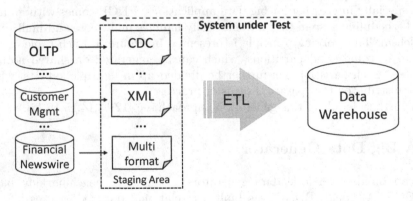

Fig. 5. TPC-DI Overview

Another recently finished data generator built based on PDGF is TPC-DI's data generator. TPC-DI is an data integration benchmark, which benchmarks ETL systems. As is shown in Figure 5, the benchmark defines several sources of information that are stored in different formats. The benchmark itself measures the performance of a system that integrates the different data sources into a single data warehouse. The data generator generates the historical files of each data source as well *change data captures* in daily increments. For the benchmark to produce meaningful results, the data from the different sources has to be consistent. This means, for example, that only employees with the status account managers in the human resources database manage customers' accounts and, thus, are referenced in the customer management database.

When combining the two examples above, one can create a big data generator that satisfies all characteristics of typical big data use cases. For example, the well established 3 to 5 V's, namely volume, velocity, variety, and the extensions value and veracity, can all be covered by such a generator. Parallel data generation is the only means to generate big volumes of data in timely fashion. The velocity aspect can be satisfied generally by fast generation of data as well as by fast generation of incremental updates, which ensure the characteristic of frequent data change. The variety aspect is covered with different data sources. The value extension is hinting to the additional value that can be retrieved from a deep analysis of the data, which therefore has to have meaningful patterns and correlations. Finally, veracity is of the data can be changed by introducing deltas and errors in the generation, which is present in the TPC-DI specification.

4   Related Work

There are multiple different approaches to data generation. Many current benchmarks use very primitive data that is simply based on statistical distributions. Examples are Terasort (a.k.a. Graysort) [1] and YCSB [2]. In order to get more realistic data, structured approaches to data generation have to be used. One way to get very realistic data is simulation. This can be done in an application specific way, e.g., using human browser interaction simulation with the Selenium simulator [19], or using a generic graph based approach [20,21]. Although very realistic, all simulation-based approaches are too slow for big data generation. Therefore, many benchmarks including most of the standard benchmarks have special purpose data generators that are not or only to a very small degree configurable. An example are all TPC benchmarks, with the exception of TPC-DI (based on PDGF) and to some extend TPC-DS (based on the partial configurable data generator MUDD [22]). Since the implementation of quality data generators is a tedious work, several commercial and scientific generic data generators have been developed. To ensure fast data generation these typically do not use simulation but either reread data to build correlations (e.g., [23]) or recompute referenced values (e.g., PDGF and Myriad [24]). Due to the data sizes generated and the speed of network and disk transfer rates, the computational approach is the fastest and most scalable and thus most suitable for big data generation.

5   Conclusion

The big data landscape is quickly evolving, much like the landscape of database management systems in its early stages. As a result, big data systems are heterogeneous and even for the broadly accepted Hadoop software stack there is no commonly accepted benchmark. Several proposals are currently emerging, because of the missing maturity of the big data field and the high pace of evolution, benchmarks have to evolve as well. To this end, configurable data generators are necessary to help benchmarks keep up with the development and, thus, stay relevant.

The Parallel Data Generation Framework is an ideal candidate for big data generation. In this article, we have listed characteristics that a big data generator has to fulfill and have demonstrated by example that PDGF can satisfy all requirements. A demo of PDGF is available for download[1]. A commercialized version is available from http://www.bankmark.de.

What is missing for an easy to use benchmark is a driver that starts the execution, measures the performance and calculates the metrics. This is non-trivial because there is no standard access language so far. However, relational input as generated by PDGF can be easily transformed in any other representation, which will ease the implementation of such tool chains.

PDGF is continuously improved and extended, current work focuses on data types typical in big data scenarios like text and click-streams. Initial implementations were used to implement the BigBench data generator. Other work targets scaling-up existing data sets and combining simulation-like data generation with the purely computational approach.

References

1. Gray, J.: GraySort Benchmark. Sort Benchmark Home Page, http://sortbenchmark.org
2. Cooper, B.F., Silberstein, A., Tam, E., Ramakrishnan, R., Sears, R.: Benchmarking Cloud Serving Systems with YCSB. In: SoCC, pp. 143–154 (2010)
3. Huang, S., Huang, J., Dai, J., Xie, T., Huang, B.: The HiBench Benchmark Suite: Characterization of the MapReduce-Based Data Analysis. In: ICDEW (2010)
4. Ghazal, A., Rabl, T., Hu, M., Raab, F., Poess, M., Crolotte, A., Jacobsen, H.A.: BigBench: Towards an industry standard benchmark for big data analytics. In: Proceedings of the ACM SIGMOD Conference (2013)
5. Baru, C., Bhandarkar, M., Nambiar, R., Poess, M., Rabl, T.: Benchmarking Big Data Systems and the BigData Top100 List. Big Data 1(1), 60–64 (2013)
6. Baru, C., Bhandarkar, M., Nambiar, R., Poess, M., Rabl, T.: Setting the Direction for Big Data Benchmark Standards. In: Nambiar, R., Poess, M. (eds.) TPCTC 2012. LNCS, vol. 7755, pp. 197–208. Springer, Heidelberg (2013)
7. Carey, M.J.: BDMS Performance Evaluation: Practices, Pitfalls, and Possibilities. In: Nambiar, R., Poess, M. (eds.) TPCTC 2012. LNCS, vol. 7755, pp. 108–123. Springer, Heidelberg (2013)
8. Patil, S., Polte, M., Ren, K., Tantisiriroj, W., Xiao, L., Lopez, J., Gibson, G., Fuchs, A., Rinaldi, B.: YCSB++: Benchmarking and performance debugging advanced features in scalable table stores. In: SoCC, pp. 9:1–9:14 (2011)
9. Rabl, T., Sadoghi, M., Jacobsen, H.A., Gómez-Villamor, S., Muntés-Mulero, V., Mankowskii, S.: Solving Big Data Challenges for Enterprise Application Performance Management. PVLDB 5(12), 1724–1735 (2012)
10. Wyatt, L., Caufield, B., Pol, D.: Principles for an ETL Benchmark. In: Nambiar, R., Poess, M. (eds.) TPCTC 2009. LNCS, vol. 5895, pp. 183–198. Springer, Heidelberg (2009)

---

[1] Parallel Data Generation Framework – http://www.paralleldatageneration.org

11. Marsaglia, G.: Xorshift RNGs. Journal of Statistical Software 8(14), 1–6 (2003)
12. Frank, M., Poess, M., Rabl, T.: Efficient Update Data Generation for DBMS Benchmark. In: ICPE 2012 (2012)
13. Poess, M., Rabl, T., Frank, M., Danisch, M.: A PDGF Implementation for TPC-H. In: Nambiar, R., Poess, M. (eds.) TPCTC 2011. LNCS, vol. 7144, pp. 196–212. Springer, Heidelberg (2012)
14. Rabl, T., Frank, M., Sergieh, H.M., Kosch, H.: A Data Generator for Cloud-Scale Benchmarking. In: Nambiar, R., Poess, M. (eds.) TPCTC 2010. LNCS, vol. 6417, pp. 41–56. Springer, Heidelberg (2011)
15. Rabl, T., Lang, A., Hackl, T., Sick, B., Kosch, H.: Generating Shifting Workloads to Benchmark Adaptability in Relational Database Systems. In: Nambiar, R., Poess, M. (eds.) TPCTC 2009. LNCS, vol. 5895, pp. 116–131. Springer, Heidelberg (2009)
16. Rabl, T., Poess, M.: Parallel data generation for performance analysis of large, complex RDBMS. In: DBTest 2011, p. 5 (2011)
17. Rabl, T., Poess, M., Danisch, M., Jacobsen, H.A.: Rapid Development of Data Generators Using Meta Generators in PDGF. In: DBTest 2013: Proceedings of the Sixth International Workshop on Testing Database Systems (2013)
18. Pöss, M., Nambiar, R.O., Walrath, D.: Why You Should Run TPC-DS: A Workload Analysis. In: VLDB, pp. 1138–1149 (2007)
19. Hunt, D., Inman-Semerau, L., May-Pumphrey, M.A., Sussman, N., Grandjean, P., Newhook, P., Suarez-Ordonez, S., Stewart, S., Kumar, T.: Selenium Documentation (2013), http://docs.seleniumhq.org/docs/
20. Houkjær, K., Torp, K., Wind, R.: Simple and Realistic Data Generation. In: VLDB 2006: Proceedings of the 32nd International Conference on Very Large Data Bases, VLDB Endowment, pp. 1243–1246 (2006)
21. Lin, P.J., Samadi, B., Cipolone, A., Jeske, D.R., Cox, S., Rendón, C., Holt, D., Xiao, R.: Development of a Synthetic Data Set Generator for Building and Testing Information Discovery Systems. In: ITNG 2006: Proceedings of the Third International Conference on Information Technology: New Generations, pp. 707–712. IEEE Computer Society, Washington, DC (2006)
22. Stephens, J.M., Poess, M.: MUDD: a multi-dimensional data generator. In: WOSP 2004: Proceedings of the 4th International Workshop on Software and Performance, pp. 104–109. ACM, New York (2004)
23. Bruno, N., Chaudhuri, S.: Flexible Database Generators. In: VLDB 2005: Proceedings of the 31st International Conference on Very Large Databases, VLDB Endowment, pp. 1097–1107 (2005)
24. Alexandrov, A., Tzoumas, K., Markl, V.: Myriad: Scalable and Expressive Data Generation. In: VLDB 2012 (2012)

From TPC-C to Big Data Benchmarks: A Functional Workload Model

Yanpei Chen[1], Francois Raab[2], and Randy Katz[3]

[1] Cloudera & UC Berkeley
yanpei@cloudera.com
[2] InfoSizing, Inc.
francois@sizing.com
[3] UC Berkeley
randy@eecs.berkeley.edu

Abstract. Big data systems help organizations store, manipulate, and derive value from vast amounts of data. Relational database and MapReduce are the two most prominent technologies for such systems. Organizations use them to perform complex analysis on diverse and unconventional data types with fast growing data volumes. As more big data systems are deployed, the industry faces the challenge to develop representative benchmarks that can evaluate the capabilities of competing implementations. In this position paper, we argue for building future big data benchmarks using what we call a "functional workload model". This concept draws on combined experiences from standard benchmarks, exemplified by TPC-C. The functional workload model describes the functional goals that the system must achieve, the data access patterns, the load variations over time, and the computation required to achieve the functional goals. Abstracting functional workload models from empirical studies of MapReduce deployments represents the first step towards building truly representative big data benchmarks.

1 Introduction

Big data systems represent one of the fastest growing segments of the computer industry today. They allow organizations to store, manipulate, and analyze large and rapidly growing volumes of data from diverse and unconventional sources. The exploding trade press on big data suggests that it has spread beyond early adopters to traditional industry sectors. As new products appear and vendors issue competing claims, the need emerges for an objective method to compare the applicability, efficiency, and cost of big data solutions. In other words, there is a growing need for a set of standard big data performance benchmarks.

There have been a number of attempts at constructing big data benchmarks [18, 20, 21, 23, 30]. None of them has yet gained wide recognition and usage. The field of big data performance is in a state where results from one publication to the next are not comparable and often not even closely related. This was also the case for online transaction processing (OLTP) some twenty years ago and for decision support shortly thereafter.

T. Rabl et al. (Eds.): WBDB 2012, LNCS 8163, pp. 28–43, 2014.

In this position paper, we propose and argue for the use of a formal process to develop standard big data benchmarks. This process draws on experiences from successful industry standard benchmarks. We start by summarizing the properties of a good benchmark and we select TPC-C, the standard yardstick for OLTP performance, to illustrate our proposed benchmark development process (Section 2). To that end, we present an insider's retrospective on the development of TPC-C and discuss the process that led to the creation of a fully synthetic, yet representative benchmark (Section 3). Analyzing this process allows us to formalize three key concepts of the paper — *application domains*, the *functional workload model* and *functions of abstraction*; and to discuss how they enable the construction of representative benchmarks that can translate across different types of big data systems (Section 4). We then highlight that the process of identifying MapReduce functional workload models and their functions of abstraction remains bottlenecked on empirical data and outline some of the challenges specific to big data benchmarks (Section 5). Finally, we present a vision for the development of big data benchmarks that would span multiple application domains, each rooted in documented empirical data (Section 6).

2   To Define a Big Data Benchmark

Performance measurement for computer systems is not a new topic, and benchmark properties are well studied. To explore the path that would lead to the definition of a successful big data benchmark, we begin by reviewing pertinent properties of a good benchmark.

2.1   Properties of a Good Benchmark

The criteria for a good performance benchmark has been the topic of multiple publications [22, 25, 27]. Prior work on the topic has identified the following essential properties:

- *Representative*: The benchmark should measure performance under real life environments and use metrics that are relevant to real life applications.
- *Relevant*: The benchmark should focus on measuring technologies that are relevant and prominent in the market and align themselves with an area where demand for performance information is high.
- *Portable*: The benchmark should be fair and portable to competing solutions that target the needs of the same applications.
- *Scalable*: The benchmark should be able to measure performance of systems within a wide range of scale. As technology progresses system scales and their performance capabilities tend to increase. The benchmark should be able to accommodate for that increase.
- *Verifiable*: The benchmark should prescribe repeatable measurements that produce the same results and can be independently verified.

- *Simple*: The conceptual elements of the benchmark should be reduced to a minimum and made easily understandable. The benchmark should also abstract away details that represent case-by-case configurations or system administration choices and do not affect performance.

While the above speaks of the properties that a good benchmark should display, it does not address the methodology through which such a benchmark can be constructed. In the following sections we propose such a methodology, illustrate it using a successful standard benchmark, and review how it can be applied to the construction of a big data benchmark.

2.2   Examples of Successful Benchmarks

The field of performance benchmarks is indeed crowded. But few benchmarks have reached the level of active industry standards. When it comes to benchmarks measuring complete or end-to-end systems, two organizations have dominated the market over the last two decades: SPEC and TPC.

Each organization has published a number of benchmarks with various degree of success. One criteria for success is the level at which the benchmark is being used by various organizations. While internal use is difficult to quantify, external publication of benchmark results is easy to tally and represents a clear success criteria. Looking at the most published benchmarks from TPC and SPEC reveals the following:

Table 1. Benchmark Result Publications

Benchmark	Publications
SPECjbb (2000 - 2005)	1,050
TPC-C	760
SPEC SFS	730
SPECweb (96 - 2009)	700
TPC-D/H	650

Of the above benchmarks, TPC-C and TPC-D/H were defined using a similar process of empirically driven abstractions. They can provide useful insight into the creation of a big data benchmark. To further explore and illustrate these concepts we will be examining the story of TPC-C with the goal to formalize the process at the core of its definition.

3   The Process of Building TPC-C

TPC-C is a good example of a benchmark that has had a substantial impact on technologies and systems. Understanding the origin of this long standing industry yardstick provides important clues toward the definition of a big data benchmark. In this section, we retrace the events that lead to the creation of TPC-C and present the conceptual motivation behind its design.

3.1   The Origins of TPC-C

The emergence and rapid growth of On Line Transaction Processing (OLTP) in the early eighties highlights the importance of benchmarking a specific application domain. The field of transaction processing was heating up and the need to satisfy on-line transaction requirements for fast user response times was growing rapidly. CODASYL databases supporting transactional properties were the dominant technology, a status increasingly challenged by relational databases. For instance, version 3 of the Oracle relational database, released in 1983, implemented support for the COMMIT and ROLLBACK functionalities. As competition intensified, the need emerged for an objective measure of performance. In 1985, Jim Gray led an industry-academia group of over twenty members to create a new OLTP benchmark under the name DebitCredit [13].

In the late eighties, relational databases had matured and were fast replacing the CODASYL model. The DebitCredit benchmark, and its derivatives ET1 and TP1, had become de-facto standards. Database system vendors used them to make performance claims, often raising controversies [33]. A single standard was still absent, which led to confusion about the comparability of results. In June of 1988, T. Sawyer and O. Serlin proposed to standardize DebitCredit. Later that year, O. Serlin spearheaded the creation of the Transaction Processing Performance Council (TPC) tasked with creating an industry standard version of DebitCredit [34].

Around this time, Digital Equipment Corporation (DEC) was in the process of developing a new relational database product, code name RdbStar. The development team soon recognized that a performance benchmark would be needed to assess the capabilities of early versions of the new product. DEC's European subsidiary had been conducting a vast empirical survey of database applications across France, England, Italy, Germany, Holland, Denmark and Finland. Production systems at key customer sites had been examined and local support staff interviewed. The survey sought to better understand how databases were used in the field and which features were most commonly found in production systems. Armed with this data, the RdbStar benchmark development project started with an examination of the many database benchmarks known at the time, including the Wisconsin benchmark [15], AS3AP [35] and the Set Query Benchmark [29].

The approach found to be the most representative of the European survey's findings came from an unpublished benchmark, one developed by the Microelectronics and Computer Consortium (MCC), one of the largest computer industry research and development consortia, based in Austin, TX. Researchers at MCC were working on distributed database technology [14] and had developed a simulator to test various designs. Part of the simulator involved executing OLTP functions inspired by an order processing application. The MCC benchmark was selected by DEC as the starting point for the RdbStar benchmark. Parts of the MCC benchmark were adjusted to be better aligned with the findings of the empirical survey and the resulting benchmark became known internally as Order-Entry.

In November of 1989, the TPC published its standardized end-to-end version of DebitCredit under the name TPC Benchmark A (TPC-A) [11]. TPC Benchmark B (TPC-B) [12] followed in August 1990, which represented a back-end version of TPC-A. By then, the simple transaction in DebitCredit was starting to come under fire as being too simplistic and not sufficiently exercising the features of mature database products. The TPC issued a request for proposal of a more complex OLTP benchmark. IBM submitted its RAMP-C benchmark and DEC submitted Order-Entry. The TPC selected the DEC benchmark and assigned its author, F. Raab, to lead the creation of the new standard. July 1992 saw the approval and release of the new TPC Benchmark C (TPC-C) [31].

3.2   The TPC-C Application Domain

One of the main purposes of a benchmark is to evaluate and contrast the merits of various implementations of the same set of requirements. These requirements are driven from the common elements found in the many use cases [28] that populate broad computational categories such as OLTP, decision support, OLAP, analytics, stream processing or big data. We use the term "application domain" to refer to these computational categories. Specifically, an application domain encapsulates many per-customer use cases. While each use case will likely include some rare and customer-specific computational needs, the application domain focuses on the common computational elements among many similar use cases.

The original Order-Entry benchmark from DEC included two distinct components: a set of database transactions targeting the OLTP application domain, and a set of simple and complex queries targeting the decision support application domain. The TPC adopted the transactional portion of Order-Entry for the creation of its new OLTP benchmark: TPC-C.

An important aspect of the design of the transactional portion of Order-Entry is that it did not follow the model traditionally used for implementing use cases and building business applications. To illustrate this aspect we contrast the two design models.

The design of a business application can be decomposed into four basic elements, as follows:

- *Tables*: The database tables, the layout of the rows and the correlation between tables.
- *Population*: The data that populates the tables, the distribution of values and the correlation between the values in different columns of the tables.
- *Transactions*: The units of computation against the data in the tables, the distribution of input variables and the interactions between transactions.
- *Scheduling*: The pacing and mix of transactions.

In the traditional design model, each of these elements implements part of the business functions targeted by the application. The tables would represent the business context. The population would start with a base set capturing the initial state of the business and evolve as a result of conducting daily business.

The transactions would implement the business functions. The scheduling would reflect business activity. This traditional model results in an application that is well aligned with the business details of the targeted use case. As such, it is too specific to be representative of the broader and more generic aspects that characterize a whole application domain.

In contrast, benchmarking is a synthetic activity that seeks to be representative of a whole application domain. Its sole purpose is to gather relevant performance information as it pertains to any application within the targeted domain. Being free of any real business context, the elements of such a benchmark can be abstracted from a representative cross section of the application domain's use cases.

To illustrate the concept of using abstractions to design the elements of a benchmark, we take a closer look at how this applies to transactions. The objective is to look at the compute units of multiple applications and to find repetitions or similarities. For instance, in the OLTP application domain, it is common to find user-initiated operations that involve multiple successive database transactions. While these transactions are related through the application's business semantics, they are otherwise independent from the point of view of exercising the system or measuring its performance. Consequently, they should be examined independently during the process of creating a set of abstract database transactions. Consider the following:

```
User-initiated operation
    Database Transaction T1
        Read row from table A
        Update row in table B
        Commit transaction
    Database Transaction T2
        Update row in table A
        Insert row in table C
        Commit transaction
    Database Transaction T3
        Read row from table C
        Update row in table B
        Commit transaction
```

In the above, T1 and T3 are performing similar operations, but on different tables. However, if tables A and C have sufficiently similar characteristics, T1 and T3 can be viewed as duplicates of the same abstract transaction, one that contains a "read row" followed by an "update row".

During the design of the Order-Entry benchmark, five abstract transactions were selected to encapsulate the activity most commonly found in real-life OLTP application. Such a simplification resulted in a substantial loss of specificity. However, we argue that the loss is more than outweighed by the gain in the ability to gather performance information that are relevant and applicable across a large portion of the OLTP application domain. The success of the benchmark over the last two decades appears to support this view.

4   Functions of Abstraction and Functional Workload Model

The process of constructing TPC-C illustrates two key concepts — *functions of abstraction* and the *functional workload model*. In this section, we explain what they are, and how they form a methodology for constructing benchmarks that target specific application domains while accommodating diverse system implementations.

4.1   Functions of Abstraction

The implementations of use cases within a particular application domain are made of computational functions, such as transactions, queries, or MapReduce jobs. As stated above, the design of a benchmark is only concerned with abstracting a cross-section of the most commonly found computational functions. We introduce the concept of *functions of abstraction* as a way of describing these abstracted computational functions. The intent is to capture a *functional* description of "what is being computed" at an abstract level; rather than a more concrete behavioral description of "how the computation is done".

The properties of a function of abstraction are as follows:

- *Generic*: The functional goal of the computation is described in a generic form, independent of the underlying system implementation, its software stack and the hardware behavior that results.
- *Atomic*: A group of transactions, queries, or jobs that must be executed together to serve a meaningful purpose (from a performance standpoint) should be considered as a single function of abstraction and not subdivided.
- *Unique*: Two different sets of transactions, queries, or jobs that serve the same functional goal are two realizations of the same function of abstraction.
- *Data independent*: The same function of abstraction can execute against data with different statistical properties and of different scales. Specifically, the description of the dataset acted upon is separate from the description of the function acting on the data.
- *Interdependent*: Their description includes the rules governing the interactions they have with each other.
- *Composable*: Any subset can be combined to create workloads of various levels of complexity.

TPC-C (i.e., Order-Entry) helps illustrate the concept. The benchmark is articulated around five functions of abstraction: a mid-weight read-write transaction (i.e., New-Order), a light-weight read-write transaction (i.e., Payment), a mid-weight read-only transaction (i.e., Order-Status), a batch of mid-weight read-write transactions (i.e., Delivery), and a heavy-weight read-only transaction (i.e., Stock-Level) [32]. They are specified in the semantic context, or story-line, of an order processing environment. That context, however, is entirely artificial. Its sole purpose is to allow easy description of the components.

Translating back to the earlier list, properties of functions of abstraction apply to TPC-C as follows:

- *Generic*: The functional goal is defined in terms of a set of data manipulation operations. The underlying system could be a relational database, a traditional file system, a CODASYL database, or an extension of the Apache Hadoop implementation of MapReduce that provides transactional capabilities.
- *Atomic*: Each transaction involves multiple data manipulation operations that operate as a whole.
- *Unique*: The five transactions serve five different functional goals.
- *Data independent*: The targeted data is defined separately through the distribution of values used as input variables. The data volume and schema is likewise specified separately.
- *Interdependent*: Their interactions is governed by the transactional properties of atomicity, consistency, isolation, and durability (i.e., the ACID properties).
- *Composable*: Workloads of various complexities can be created by using various combinations and mixes of the defined transactions. The TPC-C workload involves the combination of all five transactions, while the Payment transaction run by itself would become the TPC-A (i.e., Debit-Credit) workload.

Once defined, the functions of abstraction can be combined with a specified scheduling and with the definition of table structures and populations to form a functional workload model, which we explain next.

4.2 Functional Workload Model

The *functional workload model* captures in an implementation-independent (i.e., functional) manner the load that the system needs to service. This load is designed to be representative of the demands put on the system by an average use case within the application domain. The functional workload model includes three components - the functions of abstraction, their load pattern, and the data set they act upon.

The load pattern applied to the system is specified in terms of the execution frequency, distribution and arrival rate of each individual function of abstraction. In defining the load pattern, functions of abstraction can be combined to form coordinated groups with interdependencies.

The data set acted upon is specified in terms of its structure, inter-dependence between data elements, initial size and contents, and how it evolves over the course of the workload's execution.

The definition of these three components is limited to the essential functional goals of the particular application domain. The simplicity and lack of duplication that governs the definition of functions of abstraction must also be applied when specifying the load pattern and the data set that completes the functional workload model.

Again, TPC-C helps illustrate the concepts involved in the functional workload model:

- There are functions of abstractions in the form of five transactions.
- The load pattern involves a randomized arrival of transactions controlled by a weighted selection criteria and a random inter-arrival delay [32].
- There is an inter-dependence between the transactions. In particular, every New-Order will be accompanied by a Payment, and every group of ten New-Order transactions will produce one Delivery, one Order-Status, and one Stock-Level transaction [32].
- There are specified structures, inter-dependencies, contents, initial sizes, and growth rates for the data set, materialized in nine tables (i.e., Warehouse, District, Customer, History, Order, New-Order, Order-Line, Stock, and Item [31]).

In contrast, a major shortcoming of some of the recent big data *micro benchmark* proposals [6, 9, 26, 30] is the lack of any clear workload model, let alone a functional workload model as defined here. The resulting benchmarks measure system performance using one stand-alone compute unit at a time. They are lacking the functional view that is essential to benchmarking the diverse and rapidly changing big data solutions aimed at servicing emerging application domains, as we explain next.

4.3   Functional Benchmarks Essential for Big Data

We advocate the functional view for big data benchmarks, as illustrated by the *Functional Workload Model* layer in Figure 1.

The functional view enables a large range of similarly targeted systems to be compared, because such an abstraction level has been intentionally constructed to be independent of system implementation choices. In particular, the functional description of TPC-C does not preclude an OLTP system from being built on top of, say, the Hadoop distributed file system, and its performance compared against a relational database system.

The functional view also allows the benchmark to scale and evolve. This ability comes from the fact that functions of abstraction are specifically constructed to be independent of each other, and of the characteristics of the data sets they act upon. Thus, functions of abstraction can remain relatively fixed as the size of the data set is scaled. Further, as each application domain evolves, functions of abstraction can be added, deprecated, involved in a different load pattern or performed on a data sets with different characteristics. Thus, functions of abstraction form an essential part of a scalable and evolving benchmark model.

Figure 1 also shows the *Systems View* and *Physical View*. In Section 5.4, we will explain the pros and cons of these alternate approaches using some examples of early MapReduce benchmarks. We will also discuss these approaches in the context of a general purpose big data benchmark.

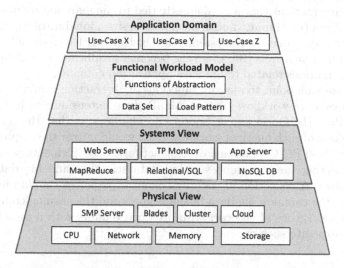

Fig. 1. The conceptual relations between application domains, functional workload models, functions of abstraction, and the system and physical views

5   Extending these Concepts to MapReduce

For the functions of abstractions concept to be useful, it must be applicable to different types of big data systems. Two important examples are relational databases and MapReduce. Identifying functions of abstraction for big data is currently bottlenecked on limited empirical knowledge. However, emerging empirical data hints toward the identification of some application domains, each with its own functional workload model. This section discusses some benchmark lessons drawn from MapReduce and generally applicable to big data.

5.1   Towards Functions of Abstraction for Big Data

MapReduce and big data represent relatively new and rapidly expanding computing paradigms. The latest empirical insights [16] indicate that the effort to extract Hadoop MapReduce functions of abstraction remains a work in progress. The data in that study, while unprecedented for MapReduce, is limited to seven workloads. This is far from the breadth of the OLTP survey that preceded TPC-C. A key result from [16] is the diversity of observed behavior. This result indicates that we should survey more system deployments to understand both common and outlier behavior. Even if functions of abstraction are extracted from the current, limited survey, there is no guarantee that these functions of abstraction would be representative of a majority of big data deployments.

A key shortcoming in the data from [16] is the lack of direct information regarding functional computation goals. This is due to the fact that current logging tools in the Apache Hadoop implementation of MapReduce collect only system-level information. Specifically, the analysis in [16] identified common MapReduce jobs using abstractions that are inherently tied to the map and reduce computational paradigm (i.e., input, shuffle, output data sizes, job durations, map and reduce task times). While such a systems-view has already led to some MapReduce-specific performance tools [10], this view becomes insufficient for extracting functions of abstractions related to big data application domains.

A good starting point to identify functions of abstraction would be to capture the data query or workflow text at MapReduce extensions such as Hive [2], Pig [4], HBase [1], Oozie [3], or Sqoop [5]. The hope is that the analysis of a large collection of such query or workflow texts would mirror the empirical survey that led to the TPC-C functions of abstraction. A complementary effort woud involve collecting the experiences of bid data scientists and big data systems administrators. A collection of such first-hand experiences should offer insights on what are the common big data business goals and the ensuing computational needs. The emergence of enterprise MapReduce vendors with a broad customer base helps expedite such efforts.

5.2 Emerging Big Data Application Domains

The data in [16] allows us to speculate on the emerging big data application domains that are addressed by the MapReduce deployments surveyed, notwithstanding the limits outlined in Section 5.1. In the following, we describe the characteristics of these application domains.

A leading application domain is *flexible latency analytics*, for which MapReduce was originally designed [19]. Flexible latency analytics is indicated by the presence of some jobs with input and output data sets that are orders of magnitude larger than for other jobs, up to the "full" data set. This application domain has previously been called "batch analytics". However, as with other application domains such as decision support, the batch nature is due to the limited capabilities of early systems. Low latency is desirable but not yet essential; hence "flexible latency". The data in [16] indicates that different deployments perform vastly different kinds of analytics, suggesting that the application domain likely involves functions of abstraction with a wide range of characteristics.

Another application domain is *interactive analytics*. Evidence suggesting interactive analytics include diurnal workload patterns, identified by visual inspection, and the presence across all workloads of frameworks such as Hive and Pig, one of whose design goals was ease of use by human analysts familiar with SQL. The presence of this application domain is confirmed by data scientists and systems administrators [8]. Low computational latency would be a major requirement. It is likely that this application domain is broader than online analytical processing (OLAP), since the analytics typically involve unstructured data, and some analyses are specifically performed to explore and identify possible data schema. The functional workload model is likely to contain a dynamic mix of functions of abstraction, with a large amount of noise and burstiness overlaid on a daily diurnal pattern.

Yet another application domain is *semi-streaming analytics*. Streaming analytics describes continuous computation processes, which often update time-aggregation metrics. For MapReduce, a common substitute for truly streaming analytics is to setup automated jobs that regularly operate on recent data, e.g., compute click-rate statistics for a social network with a job every five minutes. Since "recent" data is intentionally smaller than "historical" data, we expect functions of abstraction for this application domain to run on relatively small and uniformly sized subset of data. The functional workload model is likely to involve a steady mix of these functions of abstraction.

According to the seven deployments surveyed in [16], all three application domains appear in all big data deployments. While interactive analytics carries the most weight in terms of the number of jobs, they are all good candidates for a targeted big data benchmark, provided that they are confirmed by either trace analysis or user surveys of additional big data deployments.

5.3   Challenges Highlighted by MapReduce Survey

The MapReduce survey in [16] also served to highlight properties of big data systems that represent new challenges in the development of big data benchmarks. They can be summarized as follows:

- *System diversity*: Big data systems tend to host multiple use cases from divergent application domains. Such diversity translates to significant, and sometimes mutually exclusive, variations in the design of big data systems. A good benchmark for big data needs to replicate realistic conditions across a range of application domains, and use metrics that translates across potentially divergent computational needs. Thus, it may be challenging for a big data benchmark to be representative and portable.
- *Rapid data evolution*: Big data systems use cases constantly and rapidly evolve. This reflects the innovations in business, science, and consumer behavior facilitated by knowledge extracted from big data. This change is often rooted in the underlying data set and likely outpaces the ability to develop a representative data set as part of the functional workload model. The challenge is to ensure that the benchmark keeps sufficient pace with such changes to remain relevant.
- *System and data scale*: Big data systems often involve multiple, distributed components, while big data itself often involves multiple sources of different formats. This translates to multiple ways for the system and the data to scale. Consequently, it is challenging for a big data benchmark to be truely scalable and adequately capture the multi-dimentional scaling paradigm of big data systems.
- *System complexity*: The distributed nature of big data systems also make it challenging for a big data benchmark to be simple. Any simplifications of big data systems is likely to remain fairly complex in the absolute sense. The process of simplifications will need to be supported by objective and

empirical measurements to verify that all significant performance factors are captured by the benchmark.

5.4   Surveying MapReduce-Specific Benchmarks

The success of MapReduce greatly helped raise the profile of big data. The application domains currently dominated by MapReduce should be an important part of big data benchmarks. Some MapReduce benchmarks also help highlight limited approaches for building a general purpose big data benchmark.

The list below discusses these approaches, along with the corresponding MapReduce-specific benchmarks, and why they make it hard to achieve the desirable benchmark properties summarized in Section 2.

- Not having a true functional workload model. Bechmarks in this category focus on measuring stand-alone MapReduce jobs [6, 9, 26, 30]. They are inherently limited to measuring a narrow sliver of the full range of cluster behavior, as a real life cluster hardly ever runs one job at a time or just a handful of specific jobs. This prevents the benchmark from achieving the "representative" property.
- Adopting a physical view of benchmarking. This category includes the Gridmix3 [7] Bechmark. It seeks to reproduce the exact breakdown of jobs into tasks, the exact placement of tasks on machines, and the exact scheduling of task execution. This is similar to other physical view benchmarks that reproduce CPU, memory, disk, and network activities. While useful for comparing hardware components, one cannot use physical view benchmarks to compare, for example, two MapReduce systems that have different scheduling algorithms or operate on data of different compression formats. Further, the attempt to reproduce a large amount of execution details introduces scalability issues for the benchmark execution tool [8,24]. "Portable", "scalable", and "verifiable" properties would be hard to achieve.
- Adopting a systems view of benchmarking. This view is adopted by the SWIM [10] benchmark. This approach captures system behavior at the natural, highest level semantic boundaries in the underlying system. For MapReduce, this translates to MapReduce-specific, per job characteristics such as the input and output data to the map() and reduce() functions. The systems view does allow many desirable benchmark properties to be achieved, and SWIM is already used by leading big data platform vendors. However, the systems view for MapReduce is not "portable" to other big data solutions. For example, the map() and reduce() abstractions do not directly translate to traditional RDBMS systems. Hence, the systems view is also insufficient for a general big data benchmark.

The functional view advocated in this paper specifically seeks to go beyond these limits. It aims to enable comparison between diverse styles of systems that service the same functional goals, but have different system architectures and exhibit different physical behaviors.

6 Vision for Big Data Benchmark

The concepts of functions of abstraction, functional workload model, and application domains help us develop a vision for a possible big data benchmark.

Big data encompasses many *application domains*. OLTP is one domain. If confirmed by further survey, other possible domains are OLAP, flexible latency analytics, interactive analytics, and semi-streaming analytics. There may be other application domains yet to be identified. The criteria for identifying an application domain should be that a trace-based or user-based survey indicates that the application domain is important to the big data needs of a large range of enterprises, and that sufficient empirical traces are available to allow functions of abstraction and functional workload models to be extracted.

Within each application domain, there are multiple functions of abstraction, extracted from empirical traces and defined in the fashion outlined in Section 4.1. The benchmark should include the functions of abstraction representing the common traces from across all system deployments within the application domain. What is "common" needs to be supported by empirical traces.

There is also a representative functional workload model, extracted from empirical traces and defined in the fashion outlined in Section 4.2. Each specific system deployment or application will likely include a different organization of data sets and workload arrival patterns. The benchmark should include a single representative functional workload model for each application domain, i.e., a functional workload model that is not specific to any one application, greatly simplified, and yet typical of the entire application domain. The details of this representative functional workload model need to be supported by empirical traces.

The traces and survey used to support the selection of functions of abstraction and functional workload models should be made public. Doing so allows the benchmark to establish scientific credibility, defend against charges that it is not representative of real life conditions, and align with the business needs of enterprises seeking to derive value from big data.

Good first steps toward realizing the ideas in this paper include the Big-Bench benchmark [23], which includes English descriptions of what could be expanded into functions of abstractions for some Teradata use cases, and the CH-benchmark [17], which aims to combine the OLTP and OLAP application domains.

7 Summary and Future Work

In this paper we summarized the properties of a good benchmark and highlighted the need for a formal process to build a benchmark displaying these properties. We studied the creation of TPC-C as an example of such a process and formalized it by introducing several essential concepts — application domains, functions of abstraction, and the functional workload model. We studied the results of published surveys of big data systems as a first step toward defining application domains and functions of abstractions specific to big data, with the ultimate goal of creating a set of widely accepted and frequently used big data benchmarks.

The next step in the process of building the first standard big data benchmark would be to survey additional system deployments to identify the most prominent big data application domain and within this application domain to identify the representative functional workload model and its functions of abstraction. In the future, we should also consider combining multiple big data benchmarks to represent systems that increasingly host use cases from multiple application domains.

References

1. Apache HBase, http://hbase.apache.org/
2. Apache Hive, http://hive.apache.org/
3. Apache Oozie, http://incubator.apache.org/oozie/
4. Apache Pig, http://pig.apache.org/
5. Apache Sqoop, http://sqoop.apache.org/
6. Gridmix, HADOOP-HOME/mapred/src/benchmarks/gridmix in Hadoop 0.21.0 onwards
7. Gridmix3, HADOOP-HOME/mapred/src/contrib/gridmix in Hadoop 0.21.0 onwards
8. Personal conversation with data scientists and cluster operators at Facebook
9. Sort benchmark home page, http://sortbenchmark.org/
10. SWIM - Statistical Workload Injector for MapReduce, http://github.com/SWIMProjectUCB/SWIM/wiki
11. TPC Benchmark A Standard Specification Revision 2.0 (1994), http://www.tpc.org/tpca/spec/tpca_current.pdf
12. TPC Benchmark B Standard Specification Revision 2.0 (1994), http://www.tpc.org/tpca/spec/tpcb_current.pdf
13. Anon, et al.: A measure of transaction porcessing power. Datamation (1985)
14. Belady, L., Richter, C.: The MCC Software Technology Program. SIGSOFT 10 (1985)
15. Bitton, D., DeWitt, D., Turbyfill, C.: Benchmarking database systems: A systematic approach. In: VLDB 1983 (1983)
16. Chen, Y., Alspaugh, S., Katz, R.: Interactive Analytical Processing in Big Data Systems: A Cross-Industry Study of MapReduce Workloads. In: VLDB 2012 (2012)
17. Cole, R., et al.: The mixed workload ch-benchmark. In: DBTest 2011 (2011)
18. Cooper, B., et al.: Benchmarking cloud serving systems with ycsb. In: SOCC 2010 (2010)
19. Dean, J., Ghemawat, S.: MapReduce: simplified data processing on large clusters. In: OSDI 2004 (2004)
20. Fadika, Z., et al.: Benchmarking mapreduce implementations for application usage scenarios. In: GRID 2011 (2011)
21. Ferdman, M., et al.: Clearing the clouds, a study of emerging scale-out workloads on modern hardware. In: ASPLOS 2012 (2012)
22. Ferrari, D.: Computer systems performance evaluation. Prentice-Hall (1978)
23. Ghazal, A., et al.: Bigbench: towards an industry standard benchmark for big data analytics. In: SIGMOD 2013 (2013)

24. Gowda, B.D.: HiBench: A Representative and Comprehensive Hadoop Benchmark Suite. In: et al. (eds.) Presentations of WBDB 2012. LNCS, vol. 8163, Springer, Heidelberg (2014)
25. Gray, J.: The Benchmark Handbook For Database and Transaction Processing Systems - Introduction. In: Gray, J. (ed.) The Benchmark Handbook for Database and Transaction Processing Systems. Morgan Kaufmann Publishers (1993)
26. Huang, S., et al.: The HiBench benchmark suite: Characterization of the MapReduce-based data analysis. In: ICDEW 2010 (2010)
27. Huppler, K.: The art of building a good benchmark. In: Nambiar, R., Poess, M. (eds.) TPCTC 2009. LNCS, vol. 5895, pp. 18–30. Springer, Heidelberg (2009)
28. Jacobson, I., et al.: Object-Oriented Software Engineering - A Use Case Driven Approach. Addison-Wesley (1992)
29. O'Neil, P.: A set query benchmark for large databases. In: Conference of the Computer Measurement Group 1989 (1989)
30. Pavlo, A., et al.: A comparison of approaches to large-scale data analysis. In: SIGMOD 2009 (2009)
31. Raab, F.: TPC-C - The Standard Benchmark for Online Transaction Processing (OLTP). In: Gray, J. (ed.) The Benchmark Handbook for Database and Transaction Processing Systems. Morgan Kaufmann Publishers (1993)
32. Raab, F., Kohler, W., Shah, A.: Overview of the TPC Benchmark C: The Order-Entry Benchmark, www.tpc.org/tpcc/detail.asp
33. Serlin, O.: IBM, DEC disagree on DebitCredit results. FT Systems News 63 (1988)
34. Serlin, O.: The History of DebitCredit and the TPC. In: Gray, J. (ed.) The Benchmark Handbook for Database and Transaction Processing Systems. Morgan Kaufmann Publishers (1993)
35. Turbyfill, C., Orji, C., Bitton, D.: As3ap: A comparative relational database benchmark. In: COMPCON 1989 (1989)

The Implications of Diverse Applications and Scalable Data Sets in Benchmarking Big Data Systems

Zhen Jia[1,2], Runlin Zhou[3], Chunge Zhu[3], Lei Wang[1,2], Wanling Gao[1,2],
Yingjie Shi[1], Jianfeng Zhan[1,*], and Lixin Zhang[1]

[1] State Key Laboratory Computer Architecture, Institute of Computing Technology,
Chinese Academy of Sciences, China
[2] University of Chinese Academy of Sciences, China
[3] National Computer Network Emergency Response Technical Team Coordination
Center of China
jiazhen@ncic.ac.cn, zhourunlin@cert.org.cn, jadove@163.com,
wl@ncic.ac.cn, {gaowanling,shiyingjie,zhanjianfeng,zhanglixin}@ict.ac.cn

Abstract. Now we live in an era of big data, and big data applications
are becoming more and more pervasive. How to benchmark data center
computer systems running big data applications (in short big data systems) is a hot topic. In this paper, we focus on measuring the performance
impacts of diverse applications and scalable volumes of data sets on big
data systems. For four typical data analysis applications—an important
class of big data applications, we find two major results through experiments: first, the data scale has a significant impact on the performance of
big data systems, so we must provide scalable volumes of data sets in big
data benchmarks. Second, for the four applications, even all of them use
the simple algorithms, the performance trends are different with increasing data scales, and hence we must consider not only variety of data sets
but also variety of applications in benchmarking big data systems.

Keywords: Big Data, Benchmarking, Scalable Data.

1 Introduction

In the past decades, in order to store big data and provide services, more and more
organizations around the world build data centers with scales varying from several
nodes to hundred of thousands of nodes [21]. Massive data are produced, stored,
and analyzed in real time or off line. According to the annual survey of the global
digital output by IDC, from 2005 to 2020, the digital data will grow by a factor of
300, from 130 exabytes to 40,000 exabytes. The more data we produce, the more
data center systems are deployed for running big data applications.

As researchers in both academia and industry pay great attention to innovative systems and architecture in big data systems [5] [30] [13] [19] [20] [7] [8], the

---

* Corresponding author.

T. Rabl et al. (Eds.): WBDB 2012, LNCS 8163, pp. 44–59, 2014.
© Springer-Verlag Berlin Heidelberg 2014

pressure to evaluate and compare performance and price of these systems rises [16] [6]. Benchmarks provide fair basis for comparison among different big data systems. Besides, benchmarks represent typical needs of system support from big data applications. Together with workload characterization of typical big data applications, benchmarking results can thus enable active improvements of big data systems.

In a tutorial given at HPCA 2013 [14], we stated our position on big data benchmarking: we should take an incremental approach in stead of a top-down approach because of the following four reasons: first, there are many classes of big data applications, and there is a lack of a scientific classification of different classes of big data applications. Second, even for data center workloads, there are many important application domains, e.g., search engines, social networks, though they are mature, customers, vendors, or researchers from academia or different domains of industry do not know enough to make a big data benchmark suite because of the confidential issues [9]. Third, the value of big data drives the emergence of innovative application domains, which are far from our reach. Fourth, the complexity, diversity, scale, workload churns, and rapid evolution of big data systems indicate that both customers and vendors often have incorrect or outdated assumptions about workload behaviors [9]. Recently, big data benchmarking communities make a first but important step, and Ghazal et al. present BigBench, an end-to-end big data benchmark proposal [16], whose underlying business model of BigBench is a product retailer. Although we have some insights of the big data applications [18] [11] [30], considering the challenges mentioned above, there is a long way to go.

Workload, application and data are all important for characterizing big data systems [23]. In this paper, we focus on data analysis workloads—an important class of big data application, and investigate the performance impacts of diverse applications and scalable volumes of data set in benchmarking big data systems. We choose four typical data analysis applications from a benchmark suite for big data systems [15], and use different input data sets, the scale of which ranges from Mega Byte to Tera Byte, to drive those applications. As Rajaraman explained [22], for big data applications, inferior algorithms beat better, sophisticated algorithms because of the computing overhead. The four applications we chose indeed use simple algorithms, whose computation complexities slightly vary from $O(n)$ to $O(n \times log_2 n)$. We use a user-perceived performance metric—data processed per second to depict the system processing capability.

Through experiments, we learnt that:

- For the four representative big data applications, data scale has a significant impact on the performance of big data systems, so we must provide scalable volume of data sets in big data benchmarks.
- For the four representative big data applications, the performance trends are different with increasing data scales, and hence we must consider not only the variety of data sets but also the variety of applications when benchmarking big data systems. This also implies that there is no one-fit-all application.

The remainder of the paper is organized as follows. Section 2 shows the workloads and evaluation methodology. Section 3 reports the experiment results and Section 4 gives our analysis. Section 5 discusses the implications of our observations in benchmarking big data systems. Section 6 draws conclusions and mentions the future work.

2   Evaluation Methodology

2.1   Workloads

We choose four representative Hadoop applications from BigDataBench[15] including *Sort*, *Word Count*, *Grep* and *Naive Bayes*.

Sort is a representative I/O-intensive application, which simply uses the MapReduce framework to sort records within a directory. *Word Count* is a representative CPU-intensive application, which reads text files and counts how often the words occur. *Grep* is frequently used in data mining algorithm, and it extracts matching strings from text files. *Naive Bayes* is a simple probabilistic classifier which applies the Bayes' theorem with strong (naive) independence assumptions.

In this paper, these four applications we chosen all have relatively low computational complexity. This is because that "More data usually beats better algorithms" [22]. Table 1 shows some details of the four applications.

Table 1. Details of Different Algorithms

Application	Time Complexity	Characteristics
Sort	$O(n \times log_2 n)$	Integer comparison
WordCount	O(n)	Integer comparison and calculation
Grep	O(n)	String comparison
Naive Bayes	$O(m \times n)$	Floating-point computation

2.2   Performance Metric

We adopt a user-perceived performance metric - data processed per second to reflect the system's data processing capability. For each application, the metric of data processed per second is defined as the input data size divided by the application running time. For example, the running time of *Sort* with 100 GB input data set is 2487 seconds, and then the data processed per second of *Sort* is 41.6 MB/s. For *Sort*, this metric means the application can sort 41.6 Mega Byte data per second.

In order to explain the trend of each application's processing capability, we also collect several micro-architectural and operating system level metrics. We get the micro-architectural data by using hardware performance counters. We use Perf—a profiling tool for Linux 2.6+ based systems [2], to drive the hardware performance counters collecting micro-architectural events. In addition, we access the *proc* file system to collect OS-level performance data, such as the I/O wait time. We collect all the four slave nodes data, and report the mean value.

2.3   Summary of Hadoop Job Execution [1]

The four applications are all based on Hadoop. Hadoop is a framework that allows for the distributed processing of large data sets using the Map/Reduce model [1]. A MapRedcue job consists of a map function and a reduce function, and Hadoop breaks each job into tasks. Each map task processes one input data block (typically 64 MB) and produces intermediate results. Reduce tasks deal with the list of intermediate data through the reduce functions and produce the jobs' final output [1]. Job scheduling is performed by the unique master node of Hadoop, and there are also many slave nodes which own a fixed number of map slots and reduce slots to run tasks. The master assigns tasks of the job in response to heartbeats sent by slaves, which report the number of free map and reduce slots on the slave [28]. In our experiments, we submit the Hadoop jobs one by one and use the default FIFO scheduler policy. So the tasks of each job will be queued in the master node and be executed in FIFO orders too.

2.4   Experiment Platforms

We use a 5-node cluster to run those applications. Each node has two Xeon E5645 processors equipped with 16 GB memory and 8 TB disk. For the 5-node cluster, we deploy a Hadoop environment on it (1 master and 4 slavers). The details of configuration parameters of each node are listed in Table 2.

Table 2. Details of Configurations

CPU Type	Intel ®Xeon E5645
# Cores	6 cores@2.4G
# threads	12 threads
#Sockets	2
L1 DCache	32KB, 8-way associative, 64 byte/line
L1 ICache	32KB, 4-way associative, 64 byte/line
L2 Cache	256 KB, 8-way associative, 64 byte/line
L3 Cache	12 MB, 16-way associative, 64 byte/line
Memory	32 GB , DDR3
Network	1 Gb ethernet link

The operating system is Centos 5.5 with Linux kernel 2.6.34. The Hadoop version is 1.0.2, and the java version is JDK 1.6. For each slave node, we assign 18 map slots and 18 reduce slots with 512 MB Java heap for each slot. For other Hadoop configurations, we use the default ones.

3   Evaluation Results and Analysis

3.1   Data Scale

For those four applications, we use different input data sets to drive those applications. For *Sort*, the scale of the input data sets ranges from 200 MB to 100

GB. For *Word Count* and *Grep*, the scale of the input data sets ranges from 200 MB to 1 TB, respectively. For *Naive Bayes*, the scale of input data sets ranges from 160 MB to 300 GB. In order to eliminate the experiment deviations, each experiment is performed at least two times. We report the mean values across several times experiments.

3.2   Experiments Observations

Figure 1 shows the system's data processing capability, which is the performance metric defined in section 2.2. We can find that the system has significantly different data processing capabilities when running different applications with different scale of data sets. For example, the system processing capability running *Grep* is more than 3 times than that of running *WordCount* when they both process 1 TB data set. Meanwhile, the performance metrics of big data applications are sensitive to the data scales. Even for the same application, the processing capability is significantly varied from different scales of input data sets. For example, running *Grep*, the performance of the system is 3.077 MB per second when the data scale is 200 MB, while the processing capability is up to 398.7 MB per second with 1 TB data input. The details of our findings from those experiments are described as follows.

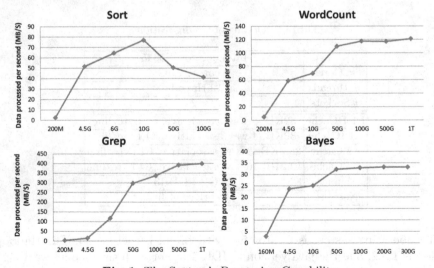

Fig. 1. The System's Processing Capability

First, different applications have different processing capabilities. We can find that the maximum processing capability is 336 MB/s (Grep) and the minimum processing capability is 33 MB/s (Naive Bayes) when the data scale is 100 GB ,respectively. This is because that *Naive Bayes* classifies records based on a probability model. It needs to calculate posterior probability for each record. So it is the most time-consuming one in the four applications, and has the lowest processing capability. While *Grep* is much simpler than the other three applications. It only finds the matched strings in each record, so it has the highest processing capability. In Section 4, we provide more performance data to explain this observation. Those different processing capabilities of different applications imply that varieties of workloads must be considered in big data benchmarking, since a certain application can not represent the behaviors of all workloads in big data field. A benchmark suite composed of diverse workloads is needed.

Second, the same application has different processing capability with different data scales. For all the four applications, we can find that there is a stage where the processing capability increases with increasing of data scale. This can be seen as a process of stressing the system step by step, which leads the system to a state of resource being fully used, and hence a peak system processing capability will appear. The reason why applications' processing capabilities increase with increasing of data scale is that the computing resources are not fully used when the data set is small, especially when the data size is less than the 4.5 GB. The basic data block size for each Hadoop map task is 64 MB in our experiments [26], so on our Hadoop cluster, the minimum data size driving all map slots to run tasks concurrently is 4.5 GB (64 $MB \times 18$ *map tasks* $\times 4$ *slaves*). When the input data set is too small (less than 4.5 GB), the Hadoop will just allocate some of map slots to complete the job. This situation causes only some of slaves busy and others less busy or even idle. So when the data set is less than 4.5 GB, applications show low processing capabilities. When all the map slots are used, the processing capabilities increase. After testing with 4.5 GB data set, we use larger data sets to stress the system further. We can find there is a turning point, of which processing capability curves stop increasing: 10 GB for *sort*, 500 GB for *Grep*, 100 GB for *WordCount* and 50 GB for *Naive Bayes*, respectively. There may be some fluctuations, which are within the range of allowable deviation. This phenomenon can be caused by many reasons, such as the different computational complexities, diverse resource requirements, and diverse system's bottlenecks, which will be further explained in Section 4. The highest points in the figure mean the maximum processing capability in our experiments. The corresponding abscissa value is the data set which can drive applications to reach the maximum processing capability. The phenomenon implies that we should tune the scalable volume of input data set to achieve the peak performance. What we must point out is that the data set size, which drives the system to reach the maximum processing capability, is an approximation for we do not enumerate all the data set size in our experiments. Take *Sort* for an example. In our experiment environment, the highest point is at 10 GB point. The input data set size, which can drive the application to reach the maximum processing capability, is about 10 GB. However, the 10 GB is an approximation, for we do not know whether a 9 GB data set or an 11 GB data set can achieve better processing capability. For the other three applications, their processing capability curves tend to smooth along with the data scale increasing. It implies that the maximum processing capabilities of them are near to the smooth points of them.

4   Further Analysis

This section will analyze the causes of phenomena in Figure 1. We will find the main factors, which cause the processing capability varying with data scale. First, we will report the cluster's resource requirements with data scale increasing, and then investigate whether the computational complexity theory can explain the processing capability trend. At last, we will explain some interesting phenomena.

4.1   Resource Requirements

As mentioned in section 3.2, increasing the input data size is a process of stressing the system and using more resources step by step. Resource consumption characteristics have great influence on the application performance [29] [25], so we would like to investigate the resource requirements and resource utilization for each application. The operation system level metrics can reflect applications' requirements directly since the operating system is the one that manages hardware resources and provides services for applications running upon it.

For an application can be decoupled into data movement and calculating, the operating system level metrics we choose are I/O wait percentage and CPU utilization, which can reflect the data movement and calculating. We get those metrics from the *proc* file system as mentioned in Section 2.2. We collect the system time, user time, irq time, softirq time and nice time, and sum those time up as the CPU used time. The CPU utilization is defined as the CPU used time divided by all CPU time. The I/O wait percentage is defined as the I/O wait time, which can also get from the *proc* file system, divided by all CPU time.

I/O wait time means the time spent by CPU waiting for I/O operations to complete. A high percentage of I/O wait time means that the application has I/O operations frequently, which further indicates that the application is an I/O intensive workload. For system, high I/O wait implies that I/O operations may be the system's bottleneck. The CPU utilization reflects how much time the CPU is used to do calculation instead of waiting for I/O or idle.

Figure 2 shows the CPU utilization and I/O wait time percentage of each workload. For *Sort*, when the data size is less than 10 GB, the data processing capability increases with the data scale increasing, and the CPU utilization goes up for it uses the Hadoop slots more efficiently. When the data size is larger than 10 GB, the processing capability decreases with data scale increasing. From Figure 2, we can find that the system's I/O wait time increases intensively whereas the CPU utilization decreases when data scale is greater than 10 GB. This phenomenon means that system is waiting for the data coming and further decreases the processing capability. The last point of *sort* application in Figure 2 seems strange. At 100 GB point the CPU usage decreases and the I/O wait time decreases at the same time, which seems unreasonable. This phenomenon is caused by the unbalanced I/O wait time [1]. The data we showed in Figure 2 is the average

---

[1] We run the *Sort* 100 GB data set several times. Each time the experiment has the similar phenomenon.

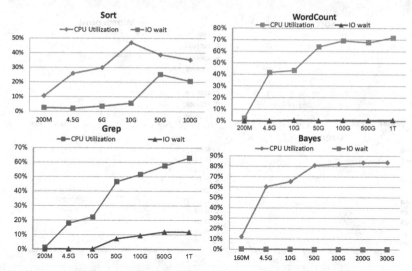

Fig. 2. The CPU Utilization and I/O Wait Percentage of Each Workloads

value of the four slaves. For the four slaves, we find that the maximum I/O wait time percentage is 31.5% and minimal I/O wait percentage is 17.6% with the average 25.3% in the face of 50 GB data. Whereas the 100GB point's maximum I/O wait percentage is 36.6% and minimal I/O wait time percentage is 10.9% with average 20.5%. The variance of running 50 GB data set is 27 whereas the variance is 94 for running 100 GB data set, which indicates that running 100 GB data set makes the system more unbalanced.

So here we can find that for the I/O intensive application – *Sort*, the processing capability trend is mostly impacted by the I/O operations. The large percentage of I/O wait time elongates the *Sort*'s execution time and further reduces the processing capability. The I/O operation becomes a bottleneck for *Sort* application.

Different from *Sort*, the other three applications (*WordCount*, *Grep* and *Naive Bayes*), are not I/O-intensive applications, and they do not have an obvious bottleneck. So the processing capability is mostly decided by CPU utilization. When the system's resource is fully used, the processing capability is unchanged.

4.2 What about Computational Complexity Theory?

The computational complexity theory is used to identify the inherent difficulty of solving a problem and it is also interested in the time consuming with an increase in the input size, which matches our scenario. The time required to solve a problem with certain scale is commonly expressed using big O notation, which is called time complexity. Such as we showed in Table 1, the time complexity of *Sort* algorithm is $O(n \times log_2 n)$. The time complexity of *Grep* and *Wordcount* is $O(n)$, and the time complexity of *Naive Bayes* is $O(m \times n)$, where m is the length of dictionary. The m is a constant, so the complexity can also be seen as $O(n)$.

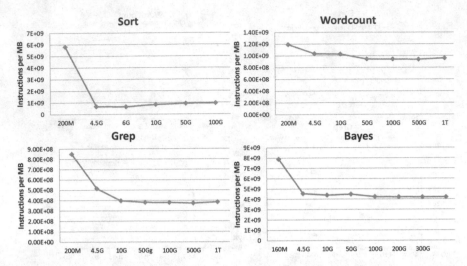

Fig. 3. Instructions Executed per Mega Byte Data Processing of Each Application

For the complexity, researchers actually use the RAM (Random Access Machine) [12] model to measure it for the Turning Machine method is incredibly tedious [17]. In order to calculate the time complexity, the researchers need to analyze the source code of the application, and find the operations in the execution path. In RAM model, each simple operation takes exactly one time step. Each memory access takes exactly one time step. Under the RAM model, the running time of an application is measured by counting up the number of time steps taking on a given input data set [24].

The RAM time-complexity is calculated by counting basic arithmetic operations in source code. And the compiler will compile the source code to instructions according to the processor's ISA (Instruction Set Architecture). The number of instructions executed can reflect how much work the processor need to do. So we collect the number of instructions executed for each workload. We calculate the instructions executed per Mega Byte data processing by using formula 1. The reason why we use this metric is that the three out of our four applications own time complexity of $O(n)$. When an algorithm's time complexity is $O(n)$, the number of instructions executed should increase in the proportion of increasing data scale. That is to say, if we double the data scale, the number of executed instructions should also be doubled. So the instructions executed per Mega Byte data should be unchanged, when the application faces different data sizes.

$$\frac{\sum_{slave1}^{slave4} Instruction\ executed}{Input\ Data\ Size\ (in\ Mega\ Byte\ )} \tag{1}$$

Figure 3 shows the number of instructions executed for processing each Mega Byte data. We can find that when the input data set is small, such as 200 MB, 4.5 GB and etc., the number of instructions executed for processing each Mega

Byte data is more than that of larger input data sets. This is because the four applications all use Hadoop framework. The framework will introduce extra instructions, such as the demon process *TaskTracker* and *DataNode* will execute many instructions. The extra instructions will affect the metric, instructions executed per Mega Byte data, especially when the input data set is small. When the data set is less than 4.5 GB, even though some slaves do not run application tasks, the Hadoop framework instructions will also be executed and counted, such as instructions executed by *DataNode* process. And hence the percentage of Hadoop framework instructions of small data set driven workload is much larger than that of workload driven by large data set. So an application driven by a small data set will execute more instructions per Mega Byte data than the large data sets. This can explain why the curves in Figure 3 decrease sharply from the smallest data set to 4.5 GB data set. When enlarging the input data set, the application will execute more application instructions and further amortize the extra instructions introduced by Hadoop framework. For the four applications, the stationary points are different. For *Grep* the stationary point is 10 GB, and for the other three applications the stationary points can be seen as 4.5 GB. Although there are some fluctuations, they are within the range of allowable deviations. The different stationary points are caused by different logic the four applications own. The *Naive Bayes* is the most complex one as mentioned in Section 3.2. So it needs the maximum number of instructions to process each Mega Byte data among the four applications (about 4.2×10^9 instructions executed for processing each Mega Byte data). The large amount of instructions needed for processing each Mega Byte data make it easy to amortize the instructions introduced by Hadoop framework. Whereas the *Grep* is the simplest one among the four applications. It needs the minimum number of instructions to process each Mega Byte data (about 3.8×10^8 instructions executed Mega Byte data). So it needs more application instructions to amortize the Hadoop framework introduced instructions. That's why the stationary point for *Grep* is 10 GB, while the points for other three are 4.5 GB. After the stationary points in Figure 3, we find the instructions executed for each Mega Byte data fit the complexity theory. The instructions executed for processing each Mega Byte data remain unchanged, when the application faces different scale data.

In Figure 3, the *Sort* has the same trend with other three applications, even though its time complexity is $O(n \times log_2 n)$. This can be explained by using the following interpretation. Let us assume that we enlarge the data set x times for *Sort*. The time complexity will be $x \times nlog_2(x \times n)$. The complexity increases $x + log_n x$ times. which can be explained by equation 2. The n in equation 2 is *Sort* application's record number. For *Sort*, each record size is about 10 KB on average. For a 10 GB input data set the record number is about 1 million, whereas the x is not a big number. The x is 5 when the data set increases from 10 GB to 50 GB. The $x \times log_n x$ will be a very small number. So the time complexity increases can be seen as x, which implies the total number of executed instructions increases at liner rate with data scale increasing. So the instructions executed for processing each Mega Byte data nearly remain the same, which is consist with Figure 3. We can conclude that the four applications' instructions executed situation meets the complexity theory.

$$\frac{x \times n \times log_2(x \times n)}{n \times log_2 n}$$
$$= \frac{x \times (log_2 x + log_2 n)}{log_2 n} \qquad (2)$$
$$= x + x \bullet \frac{log_2 x}{log_2 n}$$
$$= x + x \bullet log_n x$$

4.3 Additional Interesting Phenomena

Besides the above discussions, we also find some interesting phenomena, we will show the phenomena and explanations in the rest of this section.

Phenomenon 1: The *sort*'s processing capability trend decreases sharply when the data scale is larger than 10 GB in Figure 1.

Explanations: According to those applications' time complexity, the *Sort*'s processing capability should remain unchanged or decrease slightly after the resource is fully used. For the data processing capability can be evaluated as $n/(n \times log_2 n) = 1/log_2 n$. The processing capability will decrease at the speed $x \bullet log_n x$, which is a very small number just as explained above. But we find that *Sort* application's processing capability decreased sharply when data set is larger than 10 GB in Figure 1. The processing capability decrease between 10 GB data set and 50 GB data set reaches 53% (the process capability for 10 GB data and 50 GB data are 77.01 MB/s and 50.45 MB/s respectively). Whereas the instructions needed for processing each Mega Bytes data nearly remain unchanged (Figure 3) when the data set is larger than 10 GB for *Sort*. This phenomenon is caused by the RAM (Random Access Machine) model, which is used in calculation the time complexity. As mentioned above, the RAM model assumes that each simple operation takes exactly one time step. Each memory access takes exactly one time step, and we have as much memory as we need. The RAM model is too simple, which covers up many real situations, such as division two numbers takes more time than adding two numbers in most cases, memory access times differ greatly depending on whether data sit in cache or on the disk and etc [24]. So the RAM model can not depict the time consumed accurately, especially the long latency memory access. If the data is not in main memory, it will take a long time waiting for data coming and the I/O wait time is increased. That is to say, the long latency memory access, will cause the CPU waiting for the data coming. During this time, the instructions, which are waiting for the operand, will not be executed until the data come. The instruction is delayed and further the corresponding operation will need more time to complete. For *Sort* application, the long I/O wait time elongates the instruction execution time and further decrease the processing capability. From

Figure 2, we can find that, the I/O wait time percentage increase with the data increasing. The long I/O wait time extends the instructions execution time and makes the processing capability trend deviate from the complexity trend. Just as Larry Carter found that the performance looks much closer to $O(n^5)$ instead of $O(n^3)$ when doing matrix multiply on IBM RS/6000 [27].

For the other three applications (*WordCount, Grep, Naive Bayes*), they do not have an obvious bottleneck with the data scale increasing. Although operations and memory access do not take the same time step, the average time of processing each record tends to convergence when data volume is large enough for each application. That's why those three applications' processing capability trends meet the time complexity.

Phenomenon 2: In Figure 1, different applications have different processing capabilities even though they process the same amount of data and have the same time complexity.

Explanations: The complexity theory is used to direct algorithm design, instead of evaluating the processing capability among different kinds of algorithms. The value of time complexity (big O notation expressed) is an estimated value. The big O notation expressed time complexity is said to be described asymptotically, i.e., as the input size goes to infinity. It only includes the highest order term and excludes coefficients and lower order terms. The complexity calculated as function of the size of the input. It can give a trend of time consuming with the scale increasing when face certain problem in theory. Even though the trend may deviate from real situation, it can be used to direct algorithm design. For example, when facing the same problem, such as classification, an $O(n^2)$ algorithm is worse than an $O(n)$ algorithm e.g. *Naive Bayes*, in most instances. However, for different problems, the operations mix can be different, and the number of basic operations needed for processing each unit of data is also different. For instance, when processing the same amount of data, the instructions needed for *Grep* and *WordCount* are totally different in Figure 3. So the time complexity can not be used to evaluate different kinds of algorithms.

Actually the processing capabilities are mainly decided by the instructions executed per Mega Byte data and the systems bottleneck after the system resource is fully used. For instances, when process 100 GB data, *Grep* needs 0.39 Tera instructions whereas *Naive Bayes* needs 428 Tera instructions even though they all have the computational complexity of $O(n)$. So the *Grep* has better processing capability than *Naive Bayes*. The *Sort*'s processing capability is 77.01 MB/s when it processes 10 GB data, whereas it is 50.45 MB/s when facing 50 GB data. This is because that the percentage of I/O wait time is enlarged and becomes a bottleneck.

Phenomenon 3: Different applications' highest processing capability appears at different data scales in Figure 1.

Explanations: Different applications have different resource requirements. *Naive Bayes* needs more CPU resources than *WordCount*. When they both process 10 GB data set, *WordCount*'s CPU utilization is 43.94% whereas the *Naive Bayes*'s is 65.23% (in Figure 2). The more CPU resources needed by *Naive Bayes* drive it to reach the highest point faster. This phenomenon can explain why the definitions of "large" and "small" depend on the specific applications [10].

5   Lessons Learnt from the Experiments

Through the above experiments, we learnt several lessons in benchmarking big data systems.

5.1   Consider the Scalable Volumes of Data Inputs in Big Data Benchmarking

The data scale has a significant impact on the performance evaluation of big data systems. Even for the same application, the processing capability of the big data system in terms of data processed per second varies significantly with increasing data scales. For example, running *Grep*, the processing capability of the system is 3.077 MB per second when the data scale is 200 MB, while the processing capability is up to 398.7 MB per second with 1 TB data input. If we want to benchmark a big data system, the system should be fully used, only in this way can the system show peak performance. Big data is needed for stressing test big data systems. In addition, larger data set can reduce the impacts from framework. As mentioned in Section 4, large data set can amortize the framework introduced instructions and further decrease the framework's impacts.

From *Sort* application, we can also conclude that big data requires big data system. When we enlarge the *Sort*'s input data set, the processing capability decreases sharply for the large proportion of I/O wait time. It is too inefficient to process big data by using a small scale system. The phenomenon can also explain why more data usually beats better algorithms [22] in some degree. The big data can stress the bottleneck of the system such as I/O operations for *Sort*, so the algorithms designed for processing big data should pay more attention to avoid system's bottleneck instead of reducing the time complexity only.

In order to benchmark big data systems, we must tune the volumes of data inputs so as to get the peak performance of the system and reduce the impacts of framework, and hence scalable volumes of data input must be provided in big data benchmarks.

5.2   Consider Diversities of Workloads in Big Data Benchmarking

Also, we find that, running different applications results in varied performance number even they use the same scale of data input. For example, the processing capability of running *Grep* is more than 3 times that of running *WordCount* when they process 1 TB input data set.

As Baru et al. [7] mentioned, big data issues impinge upon a wide range of applications, covering from scientific to commercial applications. Different applications have different processing capabilities. It is difficult to single out one application to represent all. So when we evaluate big data systems, we must consider not only variety of data sets [16], but also variety of workloads. Different workloads can also reduce the impact of a specific application. Our previous work shows that customizable workloads suite is preferred to meet users' requirements [21].

5.3   The Limitation of the Sort Benchmark

Lastly, the state-of-practice methods for big data systems evaluation, such as MinuteSort[4], JouleSort, GraySort and TeraByte Sort [3], have their limitations, since most of them own a fixed scale of data input.

For example, TeraByteSort reports the performance with a 1 TB data input, which only reflects its sort performance with a 1TB data. But we do not know its performance when the data scale increases up to 10 TB or 1 PB. At the same time, we do not know whether the 1 TB data can drive the system to achieve the maximum processing capability. Another example is MinuteSort. If the MinuteSort's result of a big data system is 100 GB, which reflects that it sorts specific 100 GB data in one minute. But we do not know the processing capability in the face of 1 TB data.

Moreover, the sort benchmarks only consider one algorithm and fail to cover the diversity of workloads in big data fields.

6   Conclusion and Future Work

In this paper, we paid attention to an important class of big data applications—data analysis workloads. Through the experiments we find that first, the data scale has a significant impact on the performance of big data systems, so we must provide scalable volumes of data sets in big data benchmarks so as to achieve peak performance for big data systems with different scales. Second, for the data analysis workloads, even all of them use the simple algorithms, the performance trends are different with increasing data scales, and hence we must consider not only variety of data sets but also variety of applications in benchmarking big data systems.

For data analysis workloads, we adopt an incremental approach to build benchmark suite. Now we have investigated application domains, singled out the most important applications and released a first version benchmark suite [15] on our web page (http://prof.ict.ac.cn/BigDataBench). In the near future, we will continue to add more representative benchmarks to this suite. Especially, we will also develop data generation tools, which can generate scalable volumes of data sets for big data benchmarks.

Acknowledgment. We are very grateful to anonymous reviewers. This work is supported by the Chinese 973 project (Grant No.2011CB302502), the Hi-Tech Research and Development (863) Program of China (Grant No. 2011AA01A203, 2013AA01A213), the NSFC project (Grant No.60933003, 61202075), the BNSF project (Grant No.4133081) and the 242 project (Grant No.2012A95).

References

1. http://hadoop.apache.org/
2. Performance counters for linux,
   https://perf.wiki.kernel.org/index.php/Main_Page
3. Sort benchmark home page, http://sortbenchmark.org/
4. Apacible, J., Draves, R., et al.: Minutesort with flat datacenter storage. Technical report, Microsoft Research (2012)
5. Barroso, L., Hölzle, U.: The datacenter as a computer: An introduction to the design of warehouse-scale machines. Synthesis Lectures on Computer Architecture 4(1), 1–108 (2009)
6. Baru, C., et al.: Benchmarking big data systems and the bigdata top100 list. Big Data 1(1), 60–64 (2013)
7. Baru, C., Bhandarkar, M., Nambiar, R., Poess, M., Rabl, T.: Setting the direction for big data benchmark standards. In: Nambiar, R., Poess, M. (eds.) TPCTC 2012. LNCS, vol. 7755, pp. 197–208. Springer, Heidelberg (2013)
8. Buros, W.M., et al.: Understanding systems and architecture for big data. IBM Research Report (2013)
9. Chen, Y.: We Don't Know Enough to make a Big Data Benchmark Suite. In: Workshop on Big Data Benchmarking (2012)
10. Chen, Y., Raab, F., Katz, R.H.: From tpc-c to big data benchmarks: A functional workload model. Technical Report UCB/EECS-2012-174, EECS Department, University of California, Berkeley (July 2012)
11. Chen, Z., Jianfeng, Z., Zhen, J., Lixin, Z.: Characterizing os behavior of scale-out data center workloads. In: The Seventh Annual Workshop on the Interaction Amongst Virtualization, Operating Systems and Computer Architecture, WIVOSCA 2013 (2013)
12. Cook, S.A., Reckhow, R.A.: Time bounded random access machines. Journal of Computer and System Sciences 7(4), 354–375 (1973)
13. Ferdman, M., et al.: Clearing the clouds: A study of emerging workloads on modern hardware. Architectural Support for Programming Languages and Operating Systems (2012)
14. Gao, W., et al.: A benchmark suite for big data systems. In: The 19th IEEE International Symposium on High Performance Computer Architecture (HPCA 2013) (2013), Tutorial http://prof.ict.ac.cn/HPCA/BigDataBench.pdf
15. Gao, W., et al.: Bigdatabench: a big data benchmark suite from web search engines. In: The Third Workshop on Architectures and Systems for Big Data (ASBD 2013) in Conjunction with the 40th International Symposium on Computer Architecture (May 2013)
16. Ghazal, A., et al.: Bigbench: Towards an industry standard benchmark for big data analytics. In: ACM SIGMOD Conference (2013)
17. Holyer, I.: Computational complexity (1984)
18. Jia, Z., Wang, L., Zhan, J., Zhang, L., Luo, C.: Characterizing data analysis workloads in data centers. In: 2013 IEEE International Symposium on Workload Characterization (IISWC). IEEE (2013)
19. Jia, Z., Zhan, J., Wang, L., Zhang, L., et al.: Hvcbench: A benchmark suite for data center. The 19th IEEE International Symposium on High Performance Computer Architecture (HPCA 2013) (2013), Tutorial Technical Report http://prof.ict.ac.cn/HPCA/HPCA_Tutorial_HVC_4-jiazhen.pdf

20. Lotfi-Kamran, P., Grot, B., Ferdman, M., Volos, S., Kocberber, O., Picorel, J., Adileh, A., Jevdjic, D., Idgunji, S., Ozer, E., et al.: Scale-out processors. In: Proceedings of the 39th International Symposium on Computer Architecture, pp. 500–511. IEEE Press (2012)
21. Luo, C., Zhan, J., Jia, Z., Wang, L., Lu, G., Zhang, L., Xu, C., Sun, N.: Cloudrank-d: benchmarking and ranking cloud computing systems for data processing applications. Frontiers of Computer Science 6(4), 347–362 (2012)
22. Rajaraman, A.: More data usually beats better algorithms. Datawocky Blog (2008)
23. Sang, B., Zhan, J., Lu, G., Wang, H., Xu, D., Wang, L., Zhang, Z., Jia, Z.: Precise, scalable, and online request tracing for multitier services of black boxes. IEEE Transactions on Parallel and Distributed Systems 23(6), 1159–1167 (2012)
24. Skiena, S.S.: The algorithm design manual: with 72 figures, vol. 1. Telos Press (1998)
25. Wang, L., Zhan, J., Shi, W., Liang, Y.: In cloud, can scientific communities benefit from the economies of scale? IEEE Transactions on Parallel and Distributed Systems 23(2), 296–303 (2012)
26. White, T.: Hadoop: The definitive guide. O'Reilly Media (2012)
27. Yelick, K.: Single processor machines: Memory hierarchies and processor features
28. Zaharia, M., et al.: Delay scheduling: a simple technique for achieving locality and fairness in cluster scheduling. In: Proceedings of the 5th European Conference on Computer Systems, pp. 265–278. ACM (2010)
29. Zhan, J., Wang, L., Li, X., Shi, W., Weng, C., Zhang, W., Zang, X.: Cost-aware cooperative resource provisioning for heterogeneous workloads in data centers. IEEE Transactions on Computers
30. Zhan, J., Zhang, L., Sun, N., Wang, L., Jia, Z., Luo, C.: High volume throughput computing: Identifying and characterizing throughput oriented workloads in data centers. In: 2012 IEEE 26th International Parallel and Distributed Processing Symposium Workshops & PhD Forum (IPDPSW), pp. 1712–1721. IEEE (2012)

Processing Big Events with Showers and Streams

Christoph Doblander[1,2], Tilmann Rabl[1,3], and Hans-Arno Jacobsen[1]

[1] Middleware Systems Research Group
arno.jacobsen@msrg.org
[2] TU München, Germany
doblande@in.tum.de
[3] University of Toronto, Canada
tilmann.rabl@utoronto.ca

Abstract. Emerging use cases derived from the area of cloud computing, smart power grids, and business process management require a set of capabilities not met by traditional event processing systems. These use cases were chosen to illustrate the capabilities required from systems that are able to process what we refer to as *Big Events*, that is Big Data in motion. To further illustrate *Big Events*, we identify three use cases and analyze the characteristics of the events involved. Based on this analysis, we specify requirements regarding the event schema, event query language, historic event processing needs, event timing, and result accuracy. Collectively, we refer to the constellation of state changes in a given system that exhibits these characteristics as *event showers*, referring to the collective of these events, similar to the notion of an event stream in the context of event stream processing. We call systems that offer capabilities for meeting these requirements *event shower processing systems* in contrast to traditional *event (stream) processing systems*. The use cases we picked, demonstrate that additional value can be captured by having shower processing systems in place. The benefits lie in the new possibilities to gain additional insights, increase observability, and to further exert control and opportunities for optimizations in the given applications.

1 Introduction

As storage prices continue to drop, more and more data is stored for subsequent analysis. This trend has recently been coined as the era of *Big Data* [27]. As more and more data can be stored, the value of data analysis increases, since ever more patterns can be mined and data that previously was not of interest can be monetized. However, still many data sources are unused since their value perishes quickly. This kind of fast-paced data that needs to be processed in real-time or near-real time is often called *Big Events* [17] and refers to the processing of *Big Data* in motion.

The possibility to process *Big Events* by either exposing events from large systems or by sensing events from many sources needs a powerful processing system. We call systems that can process *Big Events*, *(event) shower processing*

T. Rabl et al. (Eds.): WBDB 2012, LNCS 8163, pp. 60–71, 2014.

systems. Events represent state transitions in the environment, conveyed as *event messages* to the system. The terms event messages and events in this context are typically used interchangeably.

More formally, an *event shower* is a partially ordered set of events, either bounded or unbounded, where the partial orderings are imposed by the causal, timing, and other relationships between the events. Others have referred to similar notions as *event clouds* [25, 26], which due to the affinity in terminology to cloud computing, we would like to avoid to reuse here. Informally speaking, an *event shower* represents the constellation of events over time resulting from considering the collective of events originating from disparate event sources in a distributed system.

Events can be sensed from the environment or can be exposed by existing systems and applications. As we show in our use case analysis, valuable information can be derived from correlating and analyzing event showers. Since our society becomes more dependent on technology and systems become more complex, observability is a critical requirement. Creating interfaces between systems requires a lot of specification and testing. In case of a malfunction, reproducing errors is a hard problem since the interactions internal to a system can not be easily reproduced or since the unique conditions leading up to the failure only occur once in a while.

Without doubt, debugging functionalities in software development tools increase observability for programmers, while event exposure increases observability in complex systems [41]. When event showers originating from multiple interacting systems can be analyzed, these systems become more observable and transparent. In case of a malfunction in an observable system, it may be possible to find ways to recover, if a well-behaved state is reached [12].

The difference between *event stream processing systems* and *event shower processing systems* are defined by the characteristics of the events involved. In stream processing systems, events tend to originate from one to a few data sources, while in shower processing systems, events originate from many data sources. Consequently, for event showers, it is impossible to accurately synchronize time across the publishing data sources. Therefore, logical clocks or other mechanism are required to establish some form of event ordering. In stream processing, events are often implicitly timestamped, relative to the single source they are emitted from or relative to the stream processor that received them in a given order, often arrival order.

In stream processing systems, the stream schema is known a priori, in *shower processing systems*, event schemas are subject to change and may not be known a priori, because the systems exposing events are not within the organizational control of the system which correlates and aggregates the events. Accordingly, *shower processing systems* have to be able to deal with schema-less information. More of the differences are discussed in Section 3.

To this end, systems are needed, which can analyze massive amounts of events from multiple systems and can deal with the characteristics of these events. With shower processing systems, it is possible to discover emergent behavior, mine for patterns, observe the interactions between disparate systems and, thus, increase the overall observability, while this is less of a concern for stream processing.

In this paper, we analyze the characteristics of events derived from considering three emerging use cases. Based on our analysis, we formulated the requirements for systems, which can deal with these kind of events, which we refer to as event shower processing system. We show how the different elements (i.e., query language, publish/subscribe semantics, and consistency requirements) fit together and outline future research required in this area in order to establish event shower processing systems. Throughout the text, we point to literature that describes systems exhibiting some of the requirements we postulate.

The rest of the paper is organized as follows. In Section 2, we describe three different use cases: cloud computing, smart power grids, and business process management. In Section 3, we define the required feature set for an event shower processing system, discuss how it differs from existing approaches, and present the individual elements of such a system. Finally, in Section 4, we discuss how event shower processing systems can be beneficial in the presented use cases.

2   Use Cases

We identified three emerging use cases where additional insights can be gained by analyzing and correlating events from multiple systems. The domains were chosen because of recent interest from the research community and where affordable sensors could bring increased observability.

Figure 1 shows an exemplary overview of a system which processes event showers. Multiple systems expose events or sense events from the environment. Operators formulate queries which compile to pattern matching rules, content-based routing topologies, aggregations and correlations. The topology is optimized for low latency or maximum throughput.

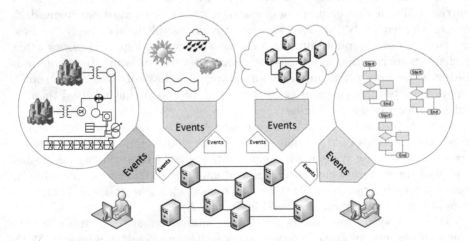

Fig. 1. Event shower processing system

2.1 Cloud Computing

With cloud computing, storage and computing resources get commoditized. Cloud providers offer on-demand configurable computing resources [28].

Monitoring plays an important role when making systems running on cloud computing platforms resilient (e.g., [39]). This helps to understand how systems operate. In case of failure, monitoring helps with the root-cause analysis or to discover potential weaknesses. As abstraction layers are added to the software stack, one looses observability because typically performance problems are understood at the very lowest layer of the stack [10].

Shower processing systems make it possible to correlate the data streams of the cloud environment with other event streams. While system-level events are well defined in terms of schema, application specific events can change frequently. A potential use case is to correlate exposed events from business processes and performance indicators of cloud environments [29]. With this information it is, for example, possible to optimize the latency of a specific business process or to forecast the impact on the infrastructure when a specific business process gets executed more often.

2.2 Smart Power Grids

The smart grid is the next evolution of the electrical power grid by enabling bidirectional communication and control of energy generators and consumers [3]. Energy demand and generation has to be exactly balanced. Until recently, the demand was given and then matched with the generation. As the portion of fluctuating energy sources, like solar and wind increases, the demand must increasingly be matched to the generation.

Observability is a key ingredient to control and balance production and demand in the smart grid [9]. This can be seen in the following example. The total production in Germany at 12:00 o'clock on June 20th, 2013, a sunny weekday, was 70 GW. 18 GW or 26% of the total production were generated by solar cells [1]. The expected solar production, published the previous day was only 13 GW, resulting in a prediction error of 5 GW. Based on the previous day's prediction, power plants are scheduled. This results in monetary losses due to severe over-provisioning. Furthermore, the gap between production and demand had to be compensated by lowering the output of conventional generators. Even more expensive is the inverse situation, where the expected solar production is not met. In this case, the higher demand has to be compensated through the spot market or more costly alternatives, with short-term regulation energy. While shifting demand is already done by huge power consumers, such as cold storages, there is a huge potential in controlling large numbers of electric vehicles and smaller devices like heat pumps or fridges [16,31,33]. Recent approaches also show how consumption forecasting could be done more accurate [44].

To control the huge number of individual devices, the current state must be observable to estimate the potential effect of control. This requires massive sensing infrastructures and near real-time processing of event showers. Since the events may have different kinds of latencies caused by changing networking conditions, the system has to deal with missing information and respect those in the overall state estimation. An area where low latency is required is, for example, phasor measurements. Phasor measurement units include GPS clocks, which provide an external time stamp for potentially correlating events, subject to the achievable GPS clock accuracy [20]. As the cost of GPS clocks and phasor measurement units decreases, it also becomes affordable to install them in low voltage grids, to pro-actively take actions in developing situations [30].

Event shower processing can add significant value to this scenario. Events from weather stations, buildings, generators, shiftable demand and energy storages can be correlated and aggregated and control can further be optimized.

2.3 Business Process Management

Exposing events from business processes can provide valuable insights if combined and correlated with external data. Instead of mining existing log files one can think of automatically exposing events via the business process execution engine [41] or to leverage a process execution engine, already designed and implemented through a publish/subscribe approach [24], thus, naturally exposing events.

Business process mining has been shown to be applicable to real-world scenarios [32]. An extended scenario could be the correlation of events from business processes with click-stream events from Web shops and weather data. An exemplary scenario is as follows: Historic click-stream events show that people tend to do more online shopping on rainy days [4]. Analysis of the events exposed from a business process engine could show that these sales have higher return rates. Thus, returns increase and customers are not satisfied. Hence, additional personnel resources are needed to deal with the returns. This shows that correlating weather data and historic events could help provision personnel accordingly.

Exposing events can also benefit white box testing in SOA environments [41]. This shows that increased observability can be used also for testing. More generally speaking, event exposure can be though of as an approach to expose unstructured information over system boundaries to enable the above described scenarios.

3 Definitions

The main characteristics of the events in the presented use cases above are the following: The events are exposed implicitly, which makes it difficult to define an event schema. They cross organizational boundaries or systems, which makes it difficult to standardize and prescribe a given event schema. Also, events may be exposed from proprietary and legacy systems, so changing the events is not easily possible. Furthermore, events from inexpensive sensors may lack exact timing information.

It is difficult to support the above use cases with existing event (stream) processing systems. While existing event processing systems exhibit capabilities to handle some of these event characteristics, event shower processing systems are representatively covering all requirements (see Table 1).

Table 1. Event showers vs. event streams

	Showers	Streams
Schema	Optional	Defined
Boundaries	Distributed across multiple systems	Part of a system
Routing	Implicit publish/subscribe semantics	-
Historic	Historic and current events	Only current events
Query language	Declarative	Can be declarative
Timing	External or logical clocks	Ordered by system arrival
Consistency	Eventual consistent	Consistent

Event Stream Processing and Complex Event Processing: Event stream processing systems typically do not cross organizational or system boundaries. If those boundaries are crossed, typically the event schema is specified, e.g., in financial markets. Complex event processing can combine multiple data sources but the correlation and aggregation of the events is done within a single system. An exemplary software which can be used in such an environment would be IBM Infosphere [2].

A system capable of processing event showers can distribute the aggregation and correlation of events across multiple systems and can consider infrastructure concerns to optimize the topology. This could be done by leveraging existing publish/subscribe-style event processing and overlay networks. [23, 43]

Rule-Based Systems: In rule-based systems it is possible to derive deductions [11]. Event shower processing systems take this approach one step further by enabling deductions on multiple event streams by supporting a declarative logic-based query language.

Publish/Subscribe Systems: Publish/Subscribe systems consist of publishers which produce events, subscribers which register for events, and brokers which route the events through an overlay network [14]. New event sources are advertised by a broadcast message. The advertisement contains schema and additional information regarding event shape and timing.

Event shower processing systems compile queries to aggregations and correlations, which are essentially join operations [22, 23]. The operations are compiled to subscriptions that attract publications as intermediate results and pass matching publications on throughout the system to higher-level subscriptions. Event shower processing systems create these implicit subscriptions to events expressed in the query language and spread the correlation and aggregation throughout the topology [22, 23].

```
:-type(production(timestamp:number(integer), household:string(varchar)),
                  watt:number(integer))).
:-type(household(hhid:string(varchar), connectedTo:string(varchar))).
:-type(solarcell(hhid:string(varchar), kwpeak:number(integer))).
:-type(windturbine(hhid:string(varchar), kwpeak:number(integer),
                   diameter:number(integer))).
:-type(transformer(trid:string(varchar), a:string(varchar),
                   b:string(varchar))).
```

Listing 1. Type definition Datalog

3.1   Event Schema

An *event schema* is a formal definition of the structure of data. Events can be
observed or are automatically exposed by systems or databases [?, 41]. When
events are exposed implicitly, the schema of these events can change, if new features are introduced in the underlying system. Hence, an event shower processing
system must be able to map unstructured, semi-structured and structured data
to a schema. An adaptor can map unstructured data to structured data, see
Figure 2 for an illustration. Existing approaches, which have been designed to
deal with semi-structured data are NoSQL databases [34]. It is also possible to
infer types based on discovered schemas. That is *type providers* [37] offer type
safe access inside a statically typed programming language. Listing 1 shows some
types from the smart grid domain. With type providers the corresponding types
and adapters could be generated automatically. This could be done based on an
advertisement, which contains type information or by discovering the schema of
events.

Fig. 2. Schema adaptor

3.2   Historic Event Data and Databases

In a system, which can process event showers, there is no difference between
current events and historic event data, see Figure 3. This is an important feature
serving the discovery and correlation between event streams or to train machine
learning models. Accessing historic event data can be implemented as a feature
of publish/subscribe systems [18, 21] or as part of a hybrid event processing
architecture, such as MADES [38], which is a distributed event store that can
query historic event data in the same way as process current and future events.

The NoSQL database CouchDB [7] can expose notifications when the underlying data changes. Relational databases have the possibility to expose events
by triggers. The notifications could be further exposed as events. This point of

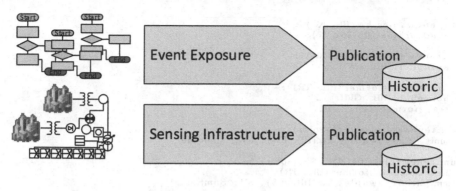

Fig. 3. Event sources

view challenges the implementation of the query processor, which should be capable to process historic data in a batch-like fashion and also incorporate current events.

Recent work [42] shows how the map-reduce model for batch processing can be combined with event stream processing. Consequently, event shower processing systems can be seen as the next evolutionary step of big data systems.

3.3 Query Language

To discover knowledge in event streams a powerful language is needed. Datalog [11] can be viewed as a subset of general logic programs. It also supports recursion which has advantages when querying graph structures, e.g., social networks or electrical grid topologies. Datalog queries are guaranteed to terminate and can be run safely [35]. It has also been demonstrated [8] that complex events can be derived from simpler events by means of deductive rules. To use Datalog for event processing some extensions are needed, for example, to reason about

```
\% transformers, transactional data
mediumhighvoltagegrid, distributiongridnorth).
transformer(munichsouth, mediumhighvoltagegrid,
distributiongridsouth). transformer(munich, highvoltagegrid,
mediumhighvoltagegrid).

\% households, transactional data
household(hh2, distributiongridsouth).
solarcell(hh2, 12).
windturbine(hh2, 23, 7).
...

\% current production, eventstreams with unix timestamp
production(1375688745, hh1, 15).
production(1375688745, hh2, 18).
production(1375688746, hh3, 4).
...
```

Listing 2. Sample data and events

```
all_renewables_kw (HH, KW) :-
   windturbine (HH, KW, _ ).

all_renewables_kw (HH, KW) :-
   solarcell (HH, KW).

nearest_transformer (HH, TR) :-
   household (HH, GRID),
   transformer (TR, _, GRID).

sum_renewables_kw (SumSolar) :-
   sum( all_renewables_kw (HH, KW), KW, SumSolar).

sum_renewables (HH, TR, SumSolar) :-
   nearest_transformer (HH, TR),
   sum( all_renewables_kw (HH, KW), KW, SumSolar).

sum_renewables (TR, SumRenewable) :-
   sum( sum_renewables (_, TR, ToSum), ToSum, SumRenewable).

current_production (TR, Prod) :-
   nearest_transformer (HH, TR),
   production (TS, HH, Prod).

sum_production (TR, SumProd) :-
   sum( current_production (TR, Prod), Prod, SumProd).
```

Listing 3. Queries in Datalog

temporal relations [5]. Also deductions and inductions have to respect temporal semantics [6]. It has already been shown that Datalog can be executed in parallel [15] and, therefore, Datalog processing can also utilize massive parallel hardware like Graphic Processing Units (GPU) or Field Programmable Gate Arrays (FPGAs). Historic data and exemplary events are shown in Listing 2. Listing 3 shows how this data is queried and aggregated. The output of the query refreshes continuously as new data becomes available or static data is updated.

3.4   Timing

Preserving ordering in a distributed system is a challenging task since clocks can not be accurately synchronized. Ordering can be preserved by logical clocks which need coordination. In case of large distributed event showers this is not practicable. Corbett et al. [13] show how to use GPS clocks to preserve ordering in a truly global distributed database. Depending on the consistency requirements approximately synchronizing to within a Δ_{max} may suffice [19]. Other approaches use heartbeat signals in streams [36]. Systems which can process event showers consequently rely on logical clocks, GPS clocks or must be aware of the synchronization error.

3.5   Accuracy

Sensors may be deployed in the field with unreliable network connections or inaccurate readings. Also, shower processing systems can span multiple organizations domains, operating world-wide, and, thus, must be self-aware of the latency they introduce. The immediate output of the event shower system may still not be accurate and as additional events arrive at the system, the result becomes more accurate. For example, if the maximum latency of incoming events from one event stream was more than a second, the result stream must be delayed at least a second to produce an accurate result.

4   Conclusions

We showed that there are emerging use cases in cloud environments, smart grids and business process management where current state of the art event processing systems are unable to cope. The events may be implicitly exposed from legacy system, business process engines or are sensed from the environment with cheap sensors and high latency network connections. Analyzing and correlating these kind of events needs additional capabilities in the query language and the underlying system, dealing, among others, with schemaless events, timing and accuracy.

We outlined possibilities to overcome those hurdles when dealing with that kind of events. First, by using a higher level logic-based query language which abstracts from publish/subscribe. Second, by adding adapters to be able to deal with unstructured events in a type-safe way. Third, by including historic data and databases which can be used to train machine learning classifiers and to discover correlations. Finally, by adding the possibility to adapt accuracy by delaying the result or by supporting logical clocks or GPS clocks.

References

1. EEX Transparency Platform (June 2013), http://transparency.eex.com
2. IBM InfoSphere (June 2013), http://www.ibm.com/software/data/infosphere/
3. NIST Smart Grid Definition (June 2013),
   http://www.nist.gov/smartgrid/beginnersguide.cfm
4. Online spending soars as shoppers stay warm in wintry conditions (June 2013),
   http://www.thisismoney.co.uk/money/markets/article-2305003
5. Allen, J.F.: Maintaining knowledge about temporal intervals. Commun. ACM 26 (November 1983)
6. Alvaro, P., Marczak, W.R., Conway, N., Hellerstein, J.M., Maier, D., Sears, R.: Dedalus: datalog in time and space. In: Proceedings of the First International Conference on Datalog Reloaded, Datalog 2010. Springer (2011)
7. Anderson, J.C., Lehnardt, J., Slater, N.: CouchDB: The Definitive Guide, 1st edn. O'Reilly Media, Inc. (2010)
8. Anicic, D., Fodor, P., Stojanovic, N., Stühmer, R.: Computing complex events in an event-driven and logic-based approach. In: Proceedings of the Third ACM International Conference on Distributed Event-Based Systems, DEBS. ACM, New York (2009)
9. Bayegan, M.: A Vision of the Future Grid. Power Engineering Review. IEEE 21(12), 10–12 (2001)

10. Cantrill, B.: Hidden in Plain Sight. ACM Queue 4(1) (February 2006)
11. Ceri, S., Gottlob, G., Tanca, L.: What you always wanted to know about datalog (and never dared to ask). IEEE Transactions on Knowledge and Data Engineering 1(1) (1989)
12. Chen, C., Ye, C., Jacobsen, H.A.: Hybrid context inconsistency resolution for context-aware services. In: IEEE International Conference on Pervasive Computing and Communications, PerCom (2011)
13. Corbett, J.C., Dean, J., Epstein, M., Fikes, A., Frost, C., Furman, J.J., Ghemawat, S., Gubarev, A., Heiser, C., Hochschild, P., Hsieh, W., Kanthak, S., Kogan, E., Li, H., Lloyd, A., Melnik, S., Mwaura, D., Nagle, D., Quinlan, S., Rao, R., Rolig, L., Saito, Y., Szymaniak, M., Taylor, C., Wang, R., Woodford, D.: Spanner: Google's globally-distributed database. In: Proceedings of the 10th USENIX Conference on Operating Systems Design and Implementation (2012)
14. Fidler, E., Jacobsen, H.A., Li, G., Mankovski, S.: The PADRES distributed publish/subscribe system. In: 8th International Conference on Feature Interactions in Telecommunications and Software Systems (2005)
15. Ganguly, S., Silberschatz, A., Tsur, S.: A framework for the parallel processing of Datalog queries. In: Proceedings of the ACM SIGMOD International Conference on Management of Data. ACM, New York (1990)
16. Goebel, C., Callaway, D.: Using ICT-Controlled Plug-in Electric Vehicles to Supply Grid Regulation in California at Different Renewable Integration Levels. IEEE Transactions on Smart Grid (2013)
17. Jacobsen, H.A.: Big events. In: Third International Workshop on Big Data Benchmarking (2013)
18. Jacobsen, H.A., Muthusamy, V., Li, G.: The PADRES Event Processing Network: Uniform Querying of Past and Future Events. IT - Information Technology 51(5), 250–260 (2009)
19. Jerzak, Z., Fach, R., Fetzer, C.: Adaptive Internal Clock Synchronization. In: IEEE Symposium on Reliable Distributed Systems, SRDS 2008 (2008)
20. Li, F., Qiao, W., Sun, H., Wan, H., Wang, J., Xia, Y., Xu, Z., Zhang, P.: Smart transmission grid: Vision and framework. IEEE Transactions on Smart Grid 1(2), 168–177 (2010)
21. Li, G., Cheung, A., Hou, S., Hu, S., Muthusamy, V., Sherafat, R., Wun, A., Jacobsen, H.A., Manovski, S.: Historic data access in publish/subscribe. In: Proceedings of the DEBS 2007. ACM, New York (2007)
22. Li, G., Jacobsen, H.-A.: Composite subscriptions in content-based publish/subscribe systems. In: Alonso, G. (ed.) Middleware 2005. LNCS, vol. 3790, pp. 249–269. Springer, Heidelberg (2005)
23. Li, G., Muthusamy, V., Jacobsen, H.-A.: Adaptive content-based routing in general overlay topologies. In: Issarny, V., Schantz, R. (eds.) Middleware 2008. LNCS, vol. 5346, pp. 1–21. Springer, Heidelberg (2008)
24. Li, G., Muthusamy, V., Jacobsen, H.A.: A distributed service-oriented architecture for business process execution. ACM Trans. Web 4(1) (January 2010)
25. Luckham, D., Schulte, R.: Event Processing Glossary - Version 2.0 (June 2013), http://www.complexevents.com/2011/08/23/event-processing-glossary-version-2-0/
26. Luckham, D.C.: The Power of Events: An Introduction to Complex Event Processing in Distributed Enterprise Systems. Addison-Wesley Longman Publishing Co., Inc., Boston (2001)

27. Manyika, J., Chui, M., Brown, B., Bughin, J., Dobbs, R., Roxburgh, C., Byers, A.H.: Big Data: The next frontier for innovation, competition, and productivity (2011), http://www.mckinsey.com/mgi/publications/big_data/pdfs/MGI_big_data_full_report.pdf

28. Mell, P., Grance, T.: NIST Cloud Computing Definition. Tech. rep. (July 2009), http://www.csrc.nist.gov/groups/SNS/cloud-computing/

29. Muthusamy, V., Jacobsen, H.-A.: BPM in Cloud Architectures: Business Process Management with SLAs and Events. In: Hull, R., Mendling, J., Tai, S. (eds.) BPM 2010. LNCS, vol. 6336, pp. 5–10. Springer, Heidelberg (2010)

30. Novosel, D., Madani, V., Bhargava, B., Vu, K., Cole, J.: Dawn of the grid synchronization. IEEE Power and Energy Magazine 6(1) (2008)

31. del Razo, V., Goebel, C.: H.A.J.: Benchmarking a car-originated-signal approach for real-time electric vehicle charging control. Tech. rep. (2013)

32. Reijers, H.A., Weijters, A.J.M.M., Dongen, B.F.V., Medeiros, A.K.A.D., Song, M., Verbeek, H.M.W.: Business process mining: An industrial application. Information Systems 32(5) (2007)

33. Rivera, J., Wolfrum, P., Hirche, S., Goebel, C., Jacobsen, H.A.: Alternating direction method of multipliers for decentralized electric vehicle charging control. In: Proceedings of the IEEE CDC (in press, 2013)

34. Sadalage, P.J., Fowler, M.: NoSQL Distilled: A Brief Guide to the Emerging World of Polyglot Persistence. Addison-Wesley Professional (2012)

35. Sagiv, Y., Vardi, M.Y.: Safety of datalog queries over infinite databases. In: Proceedings of the Eighth ACM SIGACT-SIGMOD-SIGART Symposium on Principles of Database Systems. ACM (1989)

36. Srivastava, U., Widom, J.: Flexible time management in data stream systems. In: Proceedings of the Twenty-Third ACM SIGMOD-SIGACT-SIGART Symposium on Principles of Database Systems. ACM (2004)

37. Syme, D., Battocchi, K., Takeda, K., Malayeri, D., Petricek, T.: Themes in information-rich functional programming for internet-scale data sources. In: Proceedings of the 2013 Workshop on Data Driven Functional Programming. ACM (2013)

38. Rabl, T., Mohammad Sadoghi, K.Z., Jacobsen, H.A.: DEBS 2013: Poster: MADES - A Multi-Layered, Adaptive, Distributed Event Store. ACM (2013)

39. Tseitlin, A.: The Antifragile Organization. ACM Queue (2013)

40. Ye, C., Jacobsen, H.A.: Whitening SOA Testing Via Event Exposure. IEEE Transactions on Software Engineering (2013)

41. Ye, C., Jacobsen, H.-A.: Event Exposure for Web Services: A Grey-Box Approach to Compose and Evolve Web Services. In: Chignell, M., Cordy, J., Ng, J., Yesha, Y. (eds.) The Smart Internet. LNCS, vol. 6400, pp. 197–215. Springer, Heidelberg (2010)

42. Zaharia, M., Das, T., Li, H., Shenker, S., Stoica, I.: Discretized streams: an efficient and fault-tolerant model for stream processing on large clusters. In: Proceedings of the 4th USENIX Conference on Hot Topics in Cloud Computing, HotCloud 2012, p. 10 (2012)

43. Zhang, K., Sadoghi, M., Muthusamy, V., Jacobsen, H.A.: Multicast group membership management in high speed wide area networks. In: 33rd International Conference on Distributed Computing Systems (2013)

44. Ziekow, H., Doblander, C., Goebel, C., Jacobsen, H.A.: Forecasting Household Electricity Demand with Complex Event Processing: Insights from a Prototypical Solution. In: Proceedings of the 14th International Middleware Conference. Middleware (2013)

Big Data Provenance:
Challenges and Implications for Benchmarking

Boris Glavic

Illinois Institute of Technology
10 W 31st Street, Chicago, IL 60615, USA
glavic@iit.edu

Abstract. Data Provenance is information about the origin and creation process of data. Such information is useful for debugging data and transformations, auditing, evaluating the quality of and trust in data, modelling authenticity, and implementing access control for derived data. Provenance has been studied by the database, workflow, and distributed systems communities, but provenance for Big Data - which we refer to as *Big Provenance* - is a largely unexplored field. This paper reviews existing approaches for large-scale distributed provenance and discusses potential challenges for Big Data benchmarks that aim to incorporate provenance data/management. Furthermore, we will examine how Big Data benchmarking could benefit from different types of provenance information. We argue that provenance can be used for identifying and analyzing performance bottlenecks, to compute performance metrics, and to test a system's ability to exploit commonalities in data and processing.

Keywords: Big Data, Benchmarking, Data Provenance.

1 Introduction

Provenance for Big Data applications is a relatively new topic that has not received much attention so far. A recent community white paper [5] on the challenges and opportunities of Big Data has identified provenance tracking as a major requirement for Big Data applications. Thus, provenance should be included in benchmarks targeting Big Data. We first give a brief introduction to provenance and review the current state-of-the-art of provenance for Big Data systems and applications. Afterwards, we discuss the implications of provenance for benchmarking. In particular, we try to answer the following questions: How to generate workloads with provenance aspects? What are the differences between provenance workloads and the workloads currently used in benchmarking? Finally, we argue that provenance information can be used as a supporting technology for Big Data benchmarking (for data generation and to allow new types of measurements) and profiling (enable data-centric monitoring).

2 Provenance for Big Data

Provenance information explains the creation process and origin of data by recording which transformations were responsible in creating a certain piece

T. Rabl et al. (Eds.): WBDB 2012, LNCS 8163, pp. 72–80, 2014.

of data (a so-called *data item*) and from which data items a given data item is derived. We refer to the first type as *transformation* provenance and the second type as *data provenance*. Additional meta-data such as the execution environment of a transformation (the operating system, library versions, the node that executed a transformation, ...) is sometimes also considered as provenance. A standard approach to classify provenance information is granularity. *Coarse-grained* provenance handles transformations as black-boxes: it records which data items are the inputs and outputs of a given transformation. Usually this information is represented in a graph structure by linking data items or collections to the transformations that produced or consumed them. *Fine-grained* provenance provides insights about the data-flow inside a transformation, i.e., it exposes the processing logic of a transformation by modelling which parts of the inputs were necessary/sufficient/important in deriving a specific output data item. For example, consider a transformation that counts the frequency of words in a collection of documents and outputs pairs of words and their count. If we consider documents as atomic units of data, then a coarse-grained approach would consider all input documents as the provenance of one output pair (w, c). In contrast, the fine-grained provenance of a pair (w, c) would only consist of the documents containing the word w.

Provenance has found applications in debugging data (e.g., to trace an erroneous data item back to the sources from which it was derived), trust (e.g., by combining trust scores for the data in a data item's provenance), probabilistic data (the probability of a query result can be computed from the probabilities of the data items in its provenance [13,8]), and security (e.g., enforce access-control to a query result based on access-control policies for items in its provenance [11]). All these use-cases translate to the Big Data domain. Even more, we argue that provenance is critical for applications with typical Big Data characteristics (*volume, velocity,* and *variety*)[1]. A standard approach to deal with the velocity (and to a lesser degree also the variety) aspect of Big Data is to apply data cleaning and integration steps in a pay-as-you-go fashion. This has the advantage of increasing the timeliness of data, but in comparison with the traditional ETL approach of data warehousing comes at the cost of less precise and less well-documented metadata and data transformations. Without provenance information, it is impossible for a user to understand the relevance of data, to estimate its quality, and to investigate unexpected or erroneous results. Big Data systems that automatically and transparently keep track of provenance would enable pay-as-you-go analytics that do not suffer from this loss of important metadata. Furthermore, provenance can be used to define meaningful access control policies for heavily processed and heterogenous data. For instance, a user could be granted access to analysis results if they are based on data she owns (have data that she owns in their provenance).

---

[1] To be more precise, for state-of-the-art implementations of such applications.

3   State-of-the-Art

Having motivated the need for Big provenance, we now present a brief overview of provenance research related to Big Data and highly scalable systems. Since providing a complete overview of provenance research for distributed systems is beyond the scope of this paper, we only present a few approaches that are related to Big Data research or relevant for the discussion. Provenance research from the database community has been largely focused on fine-grained provenance, but has mostly ignored distributed provenance tracking. Recently, Ikeda et al. [7] introduced an approach for tracking the provenance of workflows modelled as MapReduce jobs. The authors introduce a general fine-grained model for the provenance of map and reduce functions. Provenance is stored in HDFS by annotating each key-value pair with its provenance (appended to the value).[2] The approach provides wrappers for the map and reduce functions that call the user-provided versions of these functions. These wrappers strip off the provenance information from the value before passing it to the original user function and attach provenance to the output based on the input's provenance and the semantics of the mapper and reducer functions. The HadoopProv system [2] modifies Hadoop to achieve a similar effect. Another approach for MapReduce provenance adapts database provenance techniques to compute the provenance of workflows expressed in a subset of the Pig language [3] corresponding to relational algebra. Similarly, the approach from [15] adapts a database provenance model for a distributed datalog engine.

While most workflow systems support distributed execution of workflows, provenance techniques for these systems are mainly coarse-grained (with a few noticeable exceptions) and rely on centralized storage and processing for provenance. Malik et al. [9] present an approach for recording provenance in a distributed environment. Provenance is captured at the granularity of processes and file versions by intercepting system calls to detect dependencies between processes, files, and network connections. Each node stores parts of a provenance graph corresponding to its local processing and maintains links to the provenance graphs of other nodes. To support queries over the provenance across node boundaries, the nodes exchange summaries of their provenance graphs in the form of bloom filters. Muniswamy-Reddy et al. [10] introduce protocols for collecting provenance in a cloud environment. Each node runs *PASS* (provenance aware storage system), a system that collects provenance at the file level by intercepting system calls. Provenance is stored using cloud storage services like S3 and SimpleDB. One major concern in this work is how to guarantee that provenance and data is coupled consistently when the underlying storage services only provide eventual consistency. Seltzer et al. [12] apply the PASS approach to extend the Xen Hypervisor to collect provenance information by monitoring the system calls of a virtual machine.

---

[2] To be precise, there is an additional indirection in storing the provenance of a reducer output. See [7] for details.

In summary, existing approaches address some aspects of Big Provenance such as distributed storage, low-level operating system provenance, or fine-grained provenance for Big Data languages that are can be mapped to relational query languages (for which provenance is well-understood). However, Big Provenance still remains a challenging problem for the following reasons:

- Big data is often characterized as highly heterogeneous (*variety*) and users expect to be able to run ad-hoc analytics without having to define extensive types of meta-data like, e.g., a schema. This makes it hard to define a common structure to model the provenance of such data sets - especially for fine-grained provenance. For example, if we do not know how data entries are organized in a file, we cannot reference individual entries from the file in the provenance.
- Big Data systems tend to make the distribution of data and processing transparent to provide simpler programming models. This enables analysts with little knowledge about distributed systems to run large scale analytics. However, if the purpose of collecting provenance is to analyze the performance of a Big Data analytics system, then we would like to include information about data and processing locations in the provenance of a data item. For instance, this type of information could be used to check whether a data item was shipped to a large number of distinct locations during processing.
- A data item may have been produced by transformations that are executed using different Big Data analytics and storage solutions. The provenance of such a data item will reference data and transformations from each system that was used to create the data item. Since shipping all data items and process information in the provenance of a data item together with the data item will result in prohibitively large amounts of information to be transferred between systems, a query solution for Big Provenance has to interact with more than one system and understand several storage formats to be able to evaluate queries over provenance information.

4 Provenance as a Benchmark Workload

As mentioned before, provenance is of immense importance in the Big Data context. Thus, benchmarks for Big Data systems should include provenance workloads such as tracking provenance during the execution of a regular workload or querying pre-generated provenance data. In principle, there are two options for integrating provenance into benchmark workloads. First, existing provenance systems could be used as data generators for a benchmark and the actual workload would consist of queries over this provenance data. Second, tracking provenance could be part of the workload itself. Given the lack of Big Provenance systems discussed in Section 2, the first approach seems to be more realistic in the short term. However, in contrast to the second approach, it does not test the ability of Big Data systems to deal with provenance information. Before discussing these two options in more depth, we first discuss how provenance workloads differ from "regular" workloads and how these differences influence what aspects of a system will be stressed by a provenance workload.

4.1   Provenance vs. Standard Workloads

Typical analytics over large datasets produce outputs that are significant smaller than the input data set (e.g., clustering, outlier detection, or aggregation). Provenance, however, can be orders of magnitude larger than the data for which provenance is collected. Provenance models the relationship between inputs and outputs of a transformation and, thus, even in its simplest form, can be quadratic in the number of inputs and outputs. This increase of size is aggravated for fine-grained provenance (e.g., when tracking the provenance of each data entry in a file instead of handling the file as a single large data item) or when each data item is the result of a sequence or DAG of transformations. Furthermore, the provenance information of two data items often overlaps to a large extend [4]. A benchmark that includes workloads running over provenance data stresses a system's capability to exploits commonality in the data (e.g., compression) and to avoid unnecessary shipping of data.

4.2   Pregenerated Provenance Workloads

Because of the potential size of provenance relative to the size of the data it is describing, it is possible to generate large data sets and computationally expensive workloads by collecting the provenance of a small set of transformations at fine granularity. This property could be exploited to generate data at the scale required for benchmarking a Big Data system. A common problem with benchmark data sets of such size is that it is unfeasible to distribute full datasets effectively over the internet (limitation of network bandwidth). Hence, a Big Data benchmark should include a data generator that allows users of the benchmark to generate the data sets locally. Generating detailed provenance for a small real-world input workload using an existing provenance system is one option to realize such a data generator. In contrast to other types of data generators, this approach has the advantage that it can be bootstrapped using a small input dataset as shown in the example below.

Example 1. Consider a build process for a piece of software using the `make` build tool. During the build temporary files are created and deleted as the result of compilation. The build process executes tests on the compiled software which results in additional files being created and destroyed. Assume we execute the build using an approach that collects provenance for files by intercepting system calls (e.g., [12]). The resulting provenance graph will be large. Similarly, consider a workload that applies an image processing algorithm to an input file. We could use a provenance approach that instruments the program to record data dependencies as provenance information [14]. This would produce provenance at pixel granularity and for individual variable assignments of the image processing program. Thus, the amount of recorded provenance would be gigantic. These examples demonstrate that by computing the provenance of relatively small and simple workloads we can generate large benchmark datasets.

4.3   Provenance Tracking as Part of the Workload

Alternatively, provenance collection could be directly used as a benchmark workload. The advantage of this approach is that it measures the ability of Big Data systems to deal with provenance efficiently. However, given the current state of the art discussed in Section 4.1, a benchmark with such a workload would prevent most Big Data systems from being benchmarked. Even for systems for which provenance tracking has been realized (e.g., Hadoop) we may not want to use provenance support until its impact has been understood sufficiently well and the systems have been optimized to a reasonable extend. A solution that allows for a smoother transition is to design a workload in such a way that available provenance information could be exploited to improve performance, but is not strictly necessary to execute the workload.

Example 2. Assume a workload that requires the benchmarked system to count the appearances of words in a collection of documents (e.g., word-count for wikipedia articles from the *PUMA* benchmark [1]) and retrieve simple provenance (the original documents in which the words occur) for a small, randomly selected subset of words. A Big Data system with provenance support could use stored provenance to execute the second part of the workload efficiently while a system without provenance support could fall back to the brute force method of searching for the specific word in all documents.

In summary, the main arguments for adding provenance to Big Data benchmark workloads are:

- Provenance has been recognized as an important functionality for Big Data [5]. Thus, it is natural to expect a benchmark to test a system's capability to deal with provenance.
- Provenance workloads stress-test the ability of a system to exploit commonalities in data and processing which is essential for Big Data systems. Including provenance in a workload will allows us to generate benchmarks that target this specific aspect.

We have discussed two options for integrating provenance in benchmark workloads:

- Run an existing provenance system to pre-generate a provenance workload. Using this approach we can generate provenance benchmarks for Big Data systems without provenance support. Furthermore, the sheer size of provenance information can be exploited to (1) generate large data sets from existing small real-world workloads and (2) develop concise benchmark specifications that can be shipped and expanded to full-sized workloads locally.
- Use provenance tracking as part of workload, i.e., the benchmarked system is required to track provenance. This method would test the ability of a system to efficiently track and query provenance, but requires broad adaptation of provenance techniques for Big Data to be feasible (unless, as explained above, provenance support is made optional).

5   Data-Centric Performance Measures

Besides from being an interesting and challenging use-case for workload design, Big Provenance could also be used as a supporting technology for benchmarks. A major goal for Big Data systems is robustness of performance and scalability [6]. Provenance can be used to provide a fine-granular, data-centric view on execution times and data movement by propagating this information based on data-flow. For example, we could measure the execution times of each invocation of a mapper in a MapReduce system and attach this information as provenance to the outputs of the mapper. The individual execution times are then aggregated at the reducer and combined with the reducer's execution time. This type of provenance can be used to compute measures for individual jobs in a workload and to compute new performance metrics using provenance information.

Example 3. Assume a system performs reasonably well on a complex workload. However, one job was taking up most of the available resources while most of the jobs performed better than expected. The poor performance is hidden in the overall well performance, but may become problematic if we change the input size of the poor-performing job. We could record the execution times for all tasks of a job and the movement of data items between nodes as provenance. Based on this information we can identify jobs that use a large amount of resources relative to the size of data they consume or produce.

Note that in the example above the data-centric, provenance-based view on performance measurements is substantial for computing the measure. Benchmarks could exploit such information to define new data-centric measures for robustness of performance. For example, the benchmark could require the execution of several workloads with overlapping sets of jobs and define the deviation of execution times and data movements of a job over all workload executions as a measure of robustness.

6   Monitoring and Profiling

Acting upon the results of a benchmark to improve the performance of a system usually requires additional monitoring and profiling to identify and understand the causes of poor performance. Big Data benchmarks should consist of complex and diverse workloads. However, understanding why a system performs good or poor over a complex workload is hard. Provenance could be used to complement monitoring solutions for Big Data systems.

Assume we record resource utilization of transformations and location changes of data items as the provenance of a data item. We could compute the amount of resources that were spend on producing a data item from this type of provenance information. Note that this is the data-centric equivalent to profiling execution times of functions in, e.g., a Java program. Coupling data with performance measurements for the transformations that created it enables novel types of profiling. For example, to identify redundant computations, we simply have to check whether the provenance of the final outputs of a transformation contains a data item multiple times (possibly produced by different transformations at different locations). This information can be used to automatically detect potential optimizations (e.g., it may be cheaper to ship the data item than to reproduce it). Furthermore, if an intermediate result is not in the fine-grained provenance of any final result of a task, then it was unnecessary to produce this intermediate result at all.

7   Conclusions

This paper discusses the importance of and challenges for Big Provenance for benchmarking. In addition to sketching the advantages and issues of generating Big Data provenance workloads, we argue that provenance may also be used to aide developers in identifying bottlenecks in the performance, scalability, and robustness of their systems. Provenance can be used for 1) computing fine-grained, data-centric performance metrics, 2) for measuring if a system is able to exploit data commonalities, and 3) for profiling systems.

References

1. Ahmad, F., Lee, S., Thottethodi, M., Vijaykumar, T.: PUMA: Purdue MapReduce Benchmarks Suite. Tech. Rep. TR-ECE-12-11, Purdue University (2012)
2. Akoush, S., Sohan, R., Hopper, A.: HadoopProv: Towards Provenance as A First Class Citizen in MapReduce. TaPP (2013)
3. Amsterdamer, Y., Davidson, S., Deutch, D., Milo, T., Stoyanovich, J., Tannen, V.: Putting Lipstick on Pig: Enabling Database-style Workflow Provenance. PVLDB 5(4), 346–357 (2011)
4. Chapman, A., Jagadish, H.V., Ramanan, P.: Efficient Provenance Storage. In: SIGMOD, pp. 993–1006 (2008)
5. Divyakant, A., Bertino, E., Davidson, S., Franklin, M., Halevy, A., Han, J., Jagadish, H.V., Madden, S., Papakonstantinou, Y., Ramakrishnan, R., Ross, K., Shahabi, C., Vaithyanathan, S., Widom, J.: Challenges and opportunities with big data (2012)
6. Graefe, G.: Benchmarking robust performance. In: Rabl, T., et al. (eds.) WBDB 2012. LNCS, vol. 8163, Springer, Heidelberg (2012)
7. Ikeda, R., Park, H., Widom, J.: Provenance for generalized map and reduce workflows. In: CIDR, pp. 273–283 (2011)
8. Karvounarakis, G., Green, T.: Semiring-Annotated Data: Queries and Provenance. SIGMOD Record 41(3), 5–14 (2012)
9. Malik, T., Nistor, L., Gehani, A.: Tracking and Sketching Distributed Data Provenance. In: eScience, pp. 190–197 (2010)
10. Muniswamy-Reddy, K., Macko, P., Seltzer, M.: Provenance for the cloud. In: FAST, pp. 197–210 (2010)
11. Park, J., Nguyen, D., Sandhu, R.: A provenance-based access control model. In: PST, pp. 137–144 (2012)
12. Seltzer, M., Macko, P., Chiarini, M.: Collecting Provenance via the Xen Hypervisor. In: TaPP (2011)

13. Widom, J.: Trio: A System for Managing Data, Uncertainty, and Lineage. Managing and Mining Uncertain Data, 1–35 (2008)
14. Zhang, M., Zhang, X., Zhang, X., Prabhakar, S.: Tracing Lineage beyond Relational Operators. In: VLDB, pp. 1116–1127 (2007)
15. Zhou, W., Mapara, S., Ren, Y., Li, Y., Haeberlen, A., Ives, Z., Loo, B., Sherr, M.: Distributed time-aware provenance. PVLDB 6(2), 49–60 (2012)

Benchmarking Spatial Big Data

Shashi Shekhar[1], Michael R. Evans[1], Viswanath Gunturi[1],
KwangSoo Yang[1], and Daniel Cintra Cugler[2]

[1] Computer Science & Eng. Faculty, University of Minnesota
200 Union Street S.E. #4192, Minneapolis, MN 55455, USA
[2] Institute of Computing, University of Campinas, Campinas, SP, Brazil
{shekhar,mevans,gunturi,ksyang}@cs.umn.edu,
danielcugler@ic.unicamp.br

Abstract. Increasingly, location-aware datasets are of a size, variety,
and update rate that exceeds the capability of spatial computing technologies. This paper addresses the emerging challenges posed by such
datasets, which we call Spatial Big Data (SBD). SBD examples include trajectories of cell-phones and GPS devices, vehicle engine measurements, temporally detailed road maps, etc. SBD has the potential
to transform society via a number of new technologies including next-
generation routing services. However, the envisaged SBD-based services
pose several significant challenges for current spatial computing techniques. SBD magnifies the impact of partial information and ambiguity
of traditional routing queries specified by a start location and an end
location. In addition, SBD challenges the assumption that a single algorithm utilizing a specific dataset is appropriate for all situations. The
tremendous diversity of SBD sources substantially increases the diversity
of solution methods. Newer algorithms may emerge as new SBD becomes
available, creating the need for a flexible architecture to rapidly integrate
new datasets and associated algorithms. To quantify the performance of
these new algorithms, new benchmarks are needed that focus on these
spatial big datasets to ensure proper comparisons across techniques.

Keywords: Benchmarking, Spatial Big Data.

1 Introduction

Spatial computing is a set of ideas and technologies that facilitate understanding
the geo-physical world, knowing and communicating relations to places in that
world, and navigating through those places. The transformational potential of
mobility services is already evident. From Google Maps [17] to consumer Global
Positioning System (GPS) devices, society has benefited immensely from spatial
computing. Scientists use GPS to track endangered species to better understand
behavior, and farmers use GPS for precision agriculture to increase crop yields
while reducing costs. Google Earth is being used in classrooms to teach children
about their neighborhoods and the world in a fun and interactive way. We've
reached the point where a hiker in Yellowstone, a biker in Minneapolis, and a

T. Rabl et al. (Eds.): WBDB 2012, LNCS 8163, pp. 81–93, 2014.

taxi driver in Manhattan know precisely where they are, their nearby points of interest, and how to reach their destinations using mobility services [47].

Increasingly, however, the size, variety, and update rate of mobility datasets exceed the capacity of commonly used spatial computing and spatial database technologies to learn, manage, and process the data with reasonable effort. Such data is known as **Spatial Big Data** (SBD). We believe that harnessing SBD represents the next generation of routing services. Examples of emerging SBD datasets include temporally detailed (TD) roadmaps that provide speeds every minute for every road-segment; GPS trace data from cell-phones, and engine measurements of fuel consumption, greenhouse gas (GHG) emissions, etc. SBD has transformative potential. For example, a 2011 McKinsey Global Institute report estimates savings of "about $600 billion annually by 2020" in terms of fuel and time saved [26,29] by helping vehicles avoid congestion and reduce idling at red lights or left turns. Preliminary evidence for the transformative potential includes the experience of UPS, which saves millions of gallons of fuel by simply avoiding left turns (Figure 1(a)) and associated engine idling when selecting routes [26]. Immense savings in fuel-cost and GHG emission are possible if other fleet owners and consumers avoided left-turns and other hot spots of idling, low fuel-efficiency, and congestion. Ideas advanced in this paper may facilitate 'eco-routing' to help identify routes that reduce fuel consumption and GHG emissions, as compared to traditional route services reducing distance travelled or travel-time. It has the potential to significantly reduce US consumption of petroleum, the dominant source of energy for transportation (Figure 1(b)). It may even reduce the gap between domestic petroleum consumption and production (Figure 1(c)), helping bring the nation closer to the goal of energy independence.

A domain-specific benchmark (such as a Spatial Big Data benchmark) should address four key criteria: relevance, portability, scalability and simplicity [18]. Related work in spatial database benchmarking [36, 49] presents workloads for traditional geographic information systems (GIS) related spatial computing, such as raster and vector datasets. Raster data is used in remote sensing (e.g., Google Earth) whereas vector data represents points, lines and polygons, each with their own library of necessary operators. A key-missing component of these related benchmarks is graph-based datasets, useful for applications such as routing and urban navigation. In addition, the rise of spatio-temporal datasets also requires new workloads, as road networks now come with traffic speeds measured every minute of every day.

This paper makes the following contributions: an up-to-date workload for spatial computing, including four types of data (raster, vector, network and spatio-temporal); a set of summary metrics reminiscent of the historical TPS (transactions per second) metrics [18] for Spatial Big Data (SBD-R, -V, -N, -ST) and requirements, both functional (specific behavior) and non-functional (overall operation of a system), for future spatial computing benchmarks.

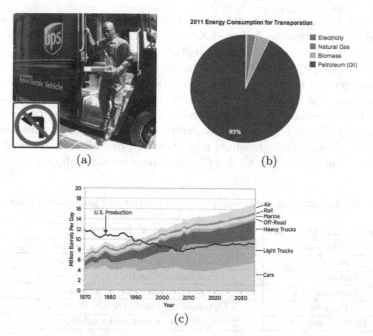

Fig. 1. (a) UPS avoids left-turns to save fuel [26]. (b) Petroleum is dominant energy source for US Transportation [54]. (c) Gap between US petroleum consumption and production is large and growing [5, 10]. (Best in color).

2   Traditional Spatial Big Data

The data inputs of spatial computing are more complex than the inputs of classical computing, as they include extended objects, such as: points, lines, and polygons in vector representation and field data in regular or irregular tessellation, such as raster data. The data inputs have two distinct types of attributes: non-spatial attributes and spatial attributes. Non-spatial attributes are used to characterize non-spatial features of objects such as name, population and unemployment rate for a city. They are the same as the attributes used in the data inputs of classical data mining. Spatial attributes are used to define the spatial location of extent of spatial objects [42]. The spatial attributes of a spatial object most often include information related to spatial locations, for example, longitude, latitude, and elevation, defined in a spatial reference frame, as well as a shape. There are four basic models to represent spatial data: raster (grid), vector (object), network (graph) and spatio-temporal:

Raster: In its simplest form, a raster consists of a matrix of cells (or pixels) organized into rows and columns (or a grid) where each cell contains a value representing information, such as temperature. A set of operations called Map Algebra was introduced [52] to manipulate representations of continuous variables defined over a common domain. These operations were categorized into three categories: local, focal and zonal; each based on the geographic size of the operation. For example, an elevation raster dataset can be queried with a zonal (large region) operation to derive slope. Raster datasets can be digital aerial photographs, imagery from satellites, digital pictures, or even scanned maps.

Vector: Geographic features in a vector format can be represented by points, lines, or polygons (areas). Vector data over a space is a framework to formalize specific relationships among a set of objects. In Table 1, a relationship between spatial and non-spatial data is described using spatial relations performed on vector datasets. These relations are separated into a number of classifications: topological to describe relationships regardless of projection, directional to describe orientation and metric to describe distances between objects.

Table 1. Common relationships among spatial and non-spatial data

Non-spatial Relation	Spatial Relation
Arithmetic	Set-oriented: union, intersection, membership, ...
Ordering	Topological: meet, within, overlap, ...
Instance-of	Directional: North, Left, Above, ...
Subclass-of	Metric: distance, area, perimeter, ...
Part-of	Dynamic: update, create, destroy, ...
Membership-of	Shape-based and visibility

Networks: Traditional spatial computing utilizes digital road maps [19, 31, 33, 46]. Figure 2(a) shows a physical road map and Figure 2(b) shows its digital, i.e., graph-based, representation. Road intersections are often modeled as vertices and the road segments connecting adjacent intersections are represented as edges in the graph. For example, the intersection of 'SE 5th Ave' and 'SE University Ave' is modeled as node N1. The segment of 'SE 5th Ave' between 'SE University Ave' and 'SE 4th Street' is represented by the edge N1-N4. The directions on the edges indicate the permitted traffic directions on the road segments. Digital roadmaps also include additional attributes for road-intersections (e.g., turn restrictions) and road-segments (e.g., centerlines, road-classification, speed-limit, historic speed, historic travel time, address-ranges, etc.) Figure 2(c) shows a tabular representation of the digital road map. Additional attributes are shown in the node and edge tables respectively. For example, the entry for edge E1 (N1-N2) in the edges table shows its speed and distance. Such datasets include roughly 100 million (10^8) edges for the roads in the U.S.A. [31].

Route determination services [28, 45], abbreviated as routing services, include the following two services: best-route determination and route comparison [41]. The first deals with determination of a best route given a start location, end location, optional waypoints, and a preference function. Here, choice of preference function could be: fastest, shortest, easiest, pedestrian, public transportation, avoid locations/areas, avoid highways, avoid toll ways, avoid U-turns, and avoid ferries. Route finding is often based on classic shortest path algorithms such as Dijktra's [24], A* [9], hierarchical [20, 21, 43, 44], materialization [38, 40, 43], and other algorithms for static graphs [4, 7, 8, 13, 14, 34, 39]. Shortest path finding is

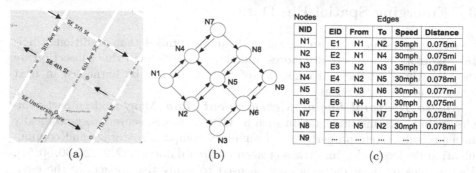

		Nodes				

Nodes	Edges				
NID	**EID**	**From**	**To**	**Speed**	**Distance**
N1	E1	N1	N2	35mph	0.075mi
N2	E2	N1	N4	30mph	0.075mi
N3	E3	N2	N3	35mph	0.078mi
N4	E4	N2	N5	30mph	0.078mi
N5	E5	N3	N6	30mph	0.077mi
N6	E6	N4	N1	30mph	0.075mi
N7	E7	N4	N7	30mph	0.078mi
N8	E8	N5	N2	30mph	0.078mi
N9	...	...	...	...	...

(a)   (b)   (c)

Fig. 2. Current representations of road maps as directed graphs with scalar travel time values. (a) Example Road Map [17]. (b) Graph Representation. (c) Tabular Representation of digital road maps.

often of interest to tourists as well as drivers in unfamiliar areas. In contrast, commuters often know a set of alternative routes between their home and work. They often use an alternate service to compare their favorite routes using real-time traffic information, e.g., scheduled maintenance and current congestion. Both services return route summary information along with auxiliary details such as route maneuver and advisory information, route geometry, route maps, and turn-by-turn instructions in an audio-visual presentation media.

OpenLS [28] presents a system (see Figure 3) that incorporates a wide-spectrum of spatial technologies, ultimately reporting to a location-aware client. The location utility performs as a geocoder by determining a geographic position, given a place name, street address or postal code. The directory service provides users with access to the nearest, or a specific place, product or service. The presentation layer renders geographic information for display. The route determination component provides routing information between locations.

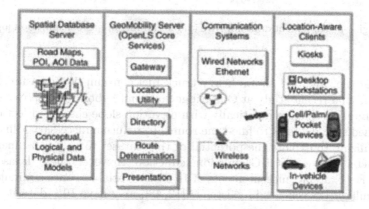

Fig. 3. OpenLS Architecture [28]

3   Emerging Spatial Big Data

Spatio-temporal datasets are significantly more detailed than traditional digital roadmaps in terms of attributes and time resolution. In this subsection we describe three representative sources of SDB that may be harnessed in next generation routing services.

Spatio-Temporal Engine Measurement Data: Many modern fleet vehicles include rich instrumentation such as GPS receivers, sensors to periodically measure sub-system properties, and auxiliary computing, storage and communication devices to log and transfer accumulated datasets [22, 23, 27, 30, 50, 51]. Engine measurement datasets may be used to study the impacts of the environment (e.g., elevation changes, weather), vehicles (e.g., weight, engine size, energy-source), traffic management systems (e.g., traffic light timing policies), and driver behaviors (e.g., gentle acceleration/braking) on fuel savings and GHG emissions.

These datasets may include a time-series of attributes such as vehicle location, fuel levels, vehicle speed, odometer values, engine speed in revolutions per minute (RPM), engine load, emissions of greenhouse gases (e.g., CO_2 and NOX), etc. Fuel efficiency can be estimated from fuel levels and distance traveled as well as engine idling from engine RPM. These attributes may be compared with geographic contexts such as elevation changes and traffic signal patterns to improve understanding of fuel efficiency and GHG emission.

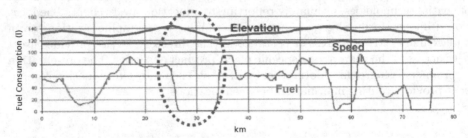

Fig. 4. Engine measurement data improve understanding of fuel consumption [6]. (Best in color).

For example, Figure 4 shows heavy truck fuel consumption as a function of elevation from a recent study at Oak Ridge National Laboratory [6]. Notice how fuel consumption changes drastically with elevation slope changes. Fleet owners have studied such datasets to fine-tune routes to reduce unnecessary idling [1, 2]. It is tantalizing to explore the potential of this dataset to help consumers gain similar fuel savings and GHG emission reduction. However, these datasets can grow big. For example, measurements of 10 engine variables, once a minute, over the 100 million US vehicles in existence [12, 48], may have 10^{14} data-items per year.

GPS Trace Data: A different type of data, GPS trajectories, is becoming available for a larger collection of vehicles due to rapid proliferation of cell-phones, in-vehicle navigation devices, and other GPS data logging devices [15,58] such as those distributed by insurance companies [57]. Such GPS traces allow indirect estimation of fuel efficiency and GHG emissions via estimation of vehicle-speed, idling and congestion. They also make it possible to make personalized route suggestions to users to reduce fuel consumption and GHG emissions. For example, Figure 5 shows 3 months of GPS trace data from a commuter with each point representing a GPS record taken at 1 minute intervals, 24 hours a day, 7 days a week. As can be seen, 3 alternative commute routes are identified between home and work from this dataset. These routes may be compared for idling, which are represented by darker (red) circles. Assuming the availability of a model to estimate fuel consumption from speed profile, one may even rank alternative routes for fuel efficiency. In recent years, consumer GPS products [15,53] are evaluating the potential of this approach.

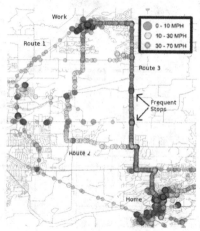

Fig. 5. A commuter's GPS tracks over three months reveal preferred routes. (Best in color).

Historical Speed Profiles: Traditionally, digital road maps consisted of centerlines and topologies of the road networks [16, 46]. These maps were used by navigation devices and web applications such as Google Maps [17] to suggest routes to users. New datasets from companies such as NAVTEQ [31] use probe vehicles and highway sensors (e.g., loop detectors) to compile travel time information across road segments for all times of the day and week at fine temporal resolutions (seconds or minutes). This data is applied to a profile model, and patterns in the road speeds are identified throughout the day. The profiles have data for every five minutes, which can then be applied to the road segment, building up an accurate picture of speeds based on historical data. Such temporally detailed (TD) roadmaps contain much more speed information than traditional roadmaps. Traditional roadmaps (Figure 2(a)) have only one scalar value of speed for any given road segment. In contrast, TD roadmaps may list speed/travel time for a road segment for thousands of time points (Figure 6) in a typical week. This allows a commuter to compare alternate start-times in addition to alternative routes. It may even allow comparison of (start-time, route) combinations to select distinct preferred routes and distinct start-times. For example, route ranking may differ across rush hour and non-rush hour and in general across different start times. However, TD roadmaps are big and their size may exceed 10^{13} items per year for the 100 million road-segments in the US when associated with per-minute values for speed or travel-time. Thus, industry is using speed-profiles, a lossy compression based on the idea of a typical day of a week, as illustrated in Figure 6(a), where each (road-segment, day of the week) pair is associated with a time-series of speed values for each hour of the day.

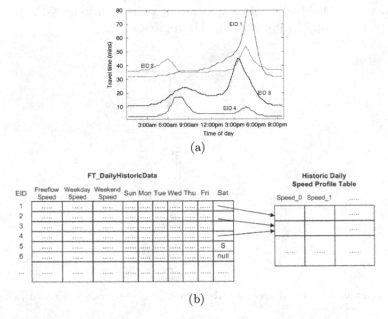

(a)

(b)

Fig. 6. Spatial Big Data on Historical Speed Profiles. (a) Travel time along four road segments over a day. (b) Schema for Daily Historic Speed Data. (Best in color).

In the near future, values for the travel time of a given edge and start time will be a distribution instead of scalar. For example, analysis of GPS tracks may show that travel-time for a road-segment is not unique, even for a given start-time of a typical week. Instead, it may consist of different values (e.g., 1, 2, 3 units), with associated frequencies (e.g., 10, 30, 20). Emergence of such SBD may allow comparison of routes, start-times and (route, start-time) combinations for statistical distribution criteria such as mean and variance. We also envision richer temporal detail on many preference functions such as fuel cost. Other emerging datasets include those related to pot-holes [35], crime reports [37], and social media reports of events on road networks [56].

4   Metrics For Spatial Big Data Benchmarks

Metrics for spatial big data can be categorized via a classification used in software engineering into functional (specific behaviors) and non-functional requirements (overall operation of a system). In this section, we will describe each and provide examples.

Metrics for Functional Spatial Big Data Requirements: Spatial computing traditionally operates on one of the four data types listed in Table 2:

Table 2. Data Types in Spatial Computing

Data Type	Representation	Operations	Potential Metric
Raster (field)	Geo-Matrix	Map algebra operations on Local, Focal, Zonal regions	Map algebra operations per second
Vector (object)	Points, Lines, Polygons	Intersection Model, Nearest Neighbor, Point Query, Range Query, etc.	Nearest Neighbors per second, Range-query (screen paint) per second
Network	Graphs (nodes, edges)	Shortest Path, Max Flow, Evacuation, etc.	Shortest-Paths per second
Spatio-Temporal	Trajectories, Temporal Networks	Time-dependent shortest path, GPS tracking, logging, etc.	Mobile device interactions per second

SBD-R: Raster datasets are frequently used for remote sensing applications, where large-scale map algebra and matrix operations are used. A helpful performance metric (e.g., map algebra operations per second) would measure how quickly representative operations of this type could be performed on a variety of dataset scales (e.g. terabyte, petabyte, exabyte, etc.).

SBD-V: Processing vector datasets in spatial database systems has historically been computationally expensive, with many key features (e.g., nearest neighbor queries) not being provided with the system. As newer systems are developed with these features, performance metrics measuring how quickly range queries and nearest neighbor queries can be computed are needed. Representative metrics include: nearest neighbors per second and range-queries per second.

SBD-N: Mapping services such as Google Maps has demonstrated the popularity of network-based datasets for use-cases such as personal transportation routing. It is not hard to imagine Google has a measure for how many shortest-paths per second it can calculate as it is serving the world driving directions, but universal and public benchmarks in this field will allow comparison between current spatial database systems. Representative metrics include: shortest-paths per second and evacuation planning.

SBD-ST: Spatio-temporal datasets are becoming more and more commonplace with the rise of location-based services and metrics for database systems rating their ability to handle some of these more common complex queries are crucial. For example, many applications currently request the location of a user, and potentially also monitor nearby points of interest to report back to the user.

So a metric that described the number of mobile device interactions (e.g., tracking, local context, location trigger, etc.) per second would be extremely useful for a variety of end-user applications. Representative metrics include: mobile device interactions per second, GPS logs per second, etc

Metrics for Non-Functional Spatial Big Data Requirements: Many Spatial Big Data use-cases (e.g., emergency services like E911, disaster response, etc.) typically require fault tolerance, where it should be resilient against natural calamities such as earthquakes, hurricanes, etc. Such requirements necessitate "triple-continental redundancy" [3], where the data is replicated on servers spread across multiple continents. This requirement poses several challenges for current cloud-based storage technologies due to performance issues inherent with wide-area replication and access. A potential metric for disaster resilience is a resilience footprint (e.g., 100 mile resilient, 1,000 mile, 10,000 mile), which may indicate the geo-spatial footprint of the disaster (e.g., fire, flood, tornado, earthquake, hurricane) that will not disrupt service.

Privacy of geographic information is an important and timely challenge due to personal information in GPS tracks, Check-in's, tweets, etc. While location information (GPS in phones and cars) can provide great value to users and industry, streams of such data also introduce spooky privacy concerns of stalking and geo-slavery [11]. For example, Ushahidi is a non-profit tech company providing technology for citizen-based reporting used in many countries with controlling regimes where privacy and protecting the reporter is extremely important [55]. Computer science efforts at obfuscating location information to date have largely yielded negative results. Thus, many individuals hesitate to indulge in mobile commerce and citizen reporting due to concern about privacy of their locations, trajectories and other spatio-temporal personal information [25]. It may be premature to provide specific metrics. However, Spatial Big Data benchmarks and metrics are needed to address many questions such as the following: "whether people reasonably expect that their movements will be recorded and aggregated..."? [32]. How do we quantify location privacy in relation to its spatio-temporal precision of measurement? How can users easily understand and set privacy constraints on location information? How does quality of location-based service change with variations in obfuscation level? Crucial to widespread adoption will be comforting the public, where a easy-to-understand metric describing the loss of privacy given information surrendered (e.g., adversary information gain per piece data submitted) will help people understand and compare various services against their privacy concerns.

5   Conclusion

Increasingly, location-aware datasets are of a size, variety, and update rate that exceed the capability of spatial computing technologies. This paper addresses the emerging challenges posed by such datasets, which we call Spatial Big Data (SBD), specifically as they apply to mobility services (e.g., transportation and routing). SBD examples include trajectories of cell-phones and GPS devices, vehicle engine measurements, temporally detailed (TD) road maps, etc. SBD has the potential to transform society.

Current benchmarks for spatial computing remain limited to small data sizes and only a portion of current popular data types. New benchmarks need to be built around Spatial Big Datasets, incorporating all four data types (raster, vector, network, spatio-temporal), while covering a wide variety of use-cases from emergency management, location-based services, advanced routing services, etc. New performance metrics, both functional (e.g., mobile interactions per second) and non-functional (e.g., disaster resilience footprint), will facilitate comparison between new systems being created and promoted by various spatial computing vendors.

Acknowledgments. We would like to thank Eric Horvitz (Microsoft), the Computing Community Consortium (CCC), Hillol Kargupta (UMBC), Erik Hoel (ESRI), Oak Ridge National Labs (ORNL) and the US-DoD for their helpful comments and support. This work was supported by NSF (0713214, DGE-0504195), US-DoD and FAPESP (grants 2011/19284-3, 2012/11395-3).

References

1. American Transportation Research Institute (ATRI). Fpm congestion monitoring at 250 freight significant highway location: Final results of the 2010 performance assessment (2010), http://goo.gl/3cAjr
2. American Transportation Research Institute (ATRI). Atri and fhwa release bottleneck analysis of 100 freight significant highway locations (2010), http://goo.gl/CONuD
3. Bauer, E., Adams, R., Eustace, D.: Beyond Redundancy: How Geographic Redundancy Can Improve Service Availability and Reliability of Computer-based Systems. Wiley-IEEE Press (2011)
4. Booth, J., Sistla, P., Wolfson, O., Cruz, I.: A data model for trip planning in multimodal transportation systems. In: Proceedings of the 12th International Conference on Extending Database Technology: Advances in Database Technology, pp. 994–1005. ACM (2009)
5. Brown, A.: Transportation Energy Futures: Addressing Key Gaps and Providing Tools for Decision Makers. Technical report, National Renewable Energy Laboratory (2011)
6. Capps, G., Franzese, O., Knee, B., Lascurain, M., Otaduy, P.: Class-8 heavy truck duty cycle project final report. ORNL/TM-2008/122 (2008)
7. Chan, E.P.F., Zhang, J.: Efficient evaluation of static and dynamic optimal route queries. In: Mamoulis, N., Seidl, T., Pedersen, T.B., Torp, K., Assent, I. (eds.) SSTD 2009. LNCS, vol. 5644, pp. 386–391. Springer, Heidelberg (2009)
8. Chang, T.: Best routes selection in international intermodal networks. Computers & Operations Research 35(9), 2877–2891 (2008)
9. Cormen, T.H., Leiserson, C.E., Rivest, R.L., Stein, C.: Introduction to Algorithms. MIT Press (2001)
10. Davis, S.C., Diegel, S.W., Boundy, R.G.: Transportation energy data book: Edition 28. Technical report, Oak Ridge National Laboratory (2010)
11. Dobson, J., Fisher, P.: Geoslavery. IEEE Technology and Society Magazine 22(1), 47–52 (2003)

12. Federal Highway Administration. Highway Statistics. HM-63, HM-64 (2008)
13. Frigioni, D., Ioffreda, M., Nanni, U., Pasqualone, G.: Experimental analysis of dynamic algorithms for the single. ACM Journal of Experimental Algorithmics (JEA) 3, 5 (1998)
14. Frigioni, D., Marchetti-Spaccamela, A., Nanni, U.: Semidynamic algorithms for maintaining single-source shortest path trees. Algorithmica 22(3), 250–274 (1998)
15. Garmin, http://www.garmin.com/us/
16. George, B., Shekhar, S.: Road maps, digital. In: Encyclopedia of GIS, pp. 967–972. Springer (2008)
17. Google Maps, http://maps.google.com
18. Gray, J.: Benchmark handbook: for database and transaction processing systems, 2nd edn. Morgan Kaufmann Publishers Inc. (1993)
19. Hoel, E.G., Heng, W.-L., Honeycutt, D.: High performance multimodal networks. In: Medeiros, C.B., Egenhofer, M., Bertino, E. (eds.) SSTD 2005. LNCS, vol. 3633, pp. 308–327. Springer, Heidelberg (2005)
20. Jagadeesh, G., Srikanthan, T., Quek, K.: Heuristic techniques for accelerating hierarchical routing on road networks. IEEE Transactions on Intelligent Transportation Systems 3(4), 301–309 (2002)
21. Jing, N., Huang, Y.-W., Rundensteiner, E.A.: Hierarchical optimization of optimal path finding for transportation applications. In: Proceedings of the Fifth International Conference on Information and Knowledge Management (CIKM), pp. 261–268. ACM (1996)
22. Kargupta, H., Gama, J., Fan, W.: The next generation of transportation systems, greenhouse emissions, and data mining. In: Proceedings of the 16th ACM SIGKDD International Conference on Knowledge Discovery and Data Mining, pp. 1209–1212. ACM (2010)
23. Kargupta, H., Puttagunta, V., Klein, M., Sarkar, K.: On-board vehicle data stream monitoring using minefleet and fast resource constrained monitoring of correlation matrices. New Generation Computing 25(1), 5–32 (2006)
24. Kleinberg, J., Tardos, E.: Algorithm Design. Pearson Education (2009)
25. Krumm, J.: A survey of computational location privacy. Personal and Ubiquitous Computing 13(6), 391–399 (2009)
26. Lovell, J.: Left-hand-turn elimination, December 9. New York Times (2007), http://goo.gl/3bkPb
27. Lynx GIS, http://www.lynxgis.com/
28. Mabrouk, M., Bychowski, T., Niedzwiadek, H., Bishr, Y., Gaillet, J., Crisp, N., Wilbrink, W., Horhammer, M., Roy, G., Margoulis, S.: Opengis location services (openls): Core services. OGC Implementation Specification 5, 016 (2005)
29. Manyika, J., et al.: Big data: The next frontier for innovation, competition and productivity. McKinsey Global Institute (May 2011)
30. MasterNaut. Green Solutions, http://www.masternaut.co.uk/carbon-calculator/
31. NAVTEQ, www.navteq.com
32. New York Times. Justices Say GPS Tracker Violated Privacy Rights (2011), http://www.nytimes.com/2012/01/24/us/police-use-of-gps-is-ruled-unconstitutional.html
33. OpenStreetMap, http://www.openstreetmap.org/
34. Potamias, M., Bonchi, F., Castillo, C., Gionis, A.: Fast shortest path distance estimation in large networks. In: Proceedings of the 18th ACM Conference on Information and Knowledge Management, CIKM 2009, pp. 867–876 (2009)

35. Pothole Info. Citizen pothole reporting via phone apps take off, but can street maintenance departments keep up? (2011), http://goo.gl/cGl3B
36. Ray, S., Simion, B., Brown, A.D.: Jackpine: A benchmark to evaluate spatial database performance. In: 2011 IEEE 27th International Conference on Data Engineering (ICDE), pp. 1139–1150. IEEE (2011)
37. SafeRoadMaps. Envisioning Safer Roads, http://saferoadmaps.org/
38. Samet, H., Sankaranarayanan, J., Alborzi, H.: Scalable network distance browsing in spatial databases. In: Proceedings of the 2008 ACM SIGMOD International Conference on Management of Data, SIGMOD 2008, pp. 43–54 (2008)
39. Sanders, P., Schultes, D.: Engineering fast route planning algorithms. In: Demetrescu, C. (ed.) WEA 2007. LNCS, vol. 4525, pp. 23–36. Springer, Heidelberg (2007)
40. Sankaranarayanan, J., Samet, H.: Query processing using distance oracles for spatial networks. IEEE Transactions on Knowledge and Data Engineering 22(8), 1158–1175 (2010)
41. Schiller, J., Voisard, A.: Location-based services. Morgan Kaufmann (2004)
42. Shekhar, S., Evans, M.R., Kang, J.M., Mohan, P.: Identifying patterns in spatial information: A survey of methods. Wiley Interdisc. Rew.: Data Mining and Knowledge Discovery 1(3), 193–214 (2011)
43. Shekhar, S., Fetterer, A., Goyal, B.: Materialization trade-offs in hierarchical shortest path algorithms. In: Scholl, M., Voisard, A. (eds.) SSD 1997. LNCS, vol. 1262, pp. 94–111. Springer, Heidelberg (1997)
44. Shekhar, S., Kohli, A., Coyle, M.: Path computation algorithms for advanced traveller information system (atis). In: Proceedings of the Ninth International Conference on Data Engineering, Vienna, Austria, April 19-23, pp. 31–39. IEEE Computer Society (1993)
45. Shekhar, S., Vatsavai, R.R., Ma, X., Yoo, J.S.: Navigation systems: A spatial database perspective. In: Location-Based Services, pp. 41–82. Morgan Kaufmann (2004)
46. Shekhar, S., Xiong, H.: Encyclopedia of GIS. Springer Publishing Company, Incorporated (2007)
47. Shrank, D., Lomax, T., Eisele, B.: The 2011 urban mobility report. Texas Transportation Institute (2011)
48. Sperling, D., Gordon, D.: Two billion cars. Oxford University Press (2009)
49. Stonebraker, M., Frew, J., Gardels, K., Meredith, J.: The sequoia 2000 storage benchmark. ACM SIGMOD Record 22, 2–11 (1993)
50. TeleNav, http://www.telenav.com/
51. TeloGIS, http://www.telogis.com/
52. Tomlin, C.D.: Geographic information systems and cartographic modeling. Prentice Hall (1990)
53. TomTom. TomTom GPS Navigation (2011), http://www.tomtom.com/
54. U.S. Energy Information Adminstration. Monthly Energy Review (June 2011), http://www.eia.gov/totalenergy/data/monthly/
55. Ushahidi, http://www.ushahidi.com
56. Waze Mobile, http://www.waze.com/
57. Wikipedia. Usage-based insurance — wikipedia, the free encyclopedia (2011), http://goo.gl/NqJE5 (accessed December 15, 2011)
58. Zhou, C., Frankowski, D., Ludford, P., Shekhar, S., Terveen, L.: Discovering personal gazetteers: an interactive clustering approach. In: Proceedings of the 12th Annual ACM International Workshop on Geographic Information Systems, pp. 266–273. ACM (2004)

Towards a Systematic Benchmark
for Array Database Systems

Peter Baumann and Heinrich Stamerjohanns

Center for Advanced Systems Engineering (CASE), Jacobs University,
Bremen, Germany
{p.baumann,h.stamerjohanns}@jacobs-university.de

Abstract. Big Data are a central challenge today in science and industry. Typically, Big Data are characterized from application perspectives. From a data structure perspective, among the core structures appearing are sets, graphs, and arrays. In particular in science and engineering we find arrays being a main contributor to data volumes. In fact, large, multi-dimensional arrays represent an important information category in earth, life, and space sciences, but also in engineering, business, and e-government.

Having long been neglected by database research, arrays today increasingly receive attention leading to a whole new field of investigation, Array Databases. As more and more Arry Database Systems emerge, similarities and differences can be observed. This calls for complementary research on benchmarks for Array DBMSs.

We present work in progress on such a comprehensive Array DBMS benchmark, which is based on our 15 years of pioneering Array DBMSs and also designing a geo raster query language standard and its corresponding functionality benchmark.

1 Motivation

Large, multi-dimensional arrays represent a major Big Data contributor in science, industry, and e-government. For example, spatio-temporal Earth science data include 1-D time series, 2-D satellite imagery, 3-D x/y/t image timeseries and x/y/z geophysical data, and 4-D x/y/z/t atmosphere and ocean data, among others. Likewise, in Life Sciences bio/medical modalities like computerized tomography (CT) scans and confocal microscopy produce increasing amounts of spatio-temporal data. In Astrophysics, optical and radar sensors deliver high-resolution raster data as continuous streams and with large numbers of spectral bands. Statistical data sets transcend spatio-temporal dimensions by using user-defined measures as dimension axes, but still yielding n-D data cubes. Multimedia databases use vectors of hundreds to thousands of features for content-based image retrieval. Figure 1 symbolizes some relevant applications.

Arrays appear as low-dimensional spatio-temporal data, medium-dimension statistics data (such as 3 to 12 dimensional OLAP [17]), and high-dimensional feature vectors (with thousands of dimensions) [15]. A further distinguishing

T. Rabl et al. (Eds.): WBDB 2012, LNCS 8163, pp. 94–102, 2014.

Fig. 1. Array data: A collage of applications

criterion is the number of cells carrying meaningful information: sparse data, with typically 3 to 5 % of cell positions being occupied by data, appear in OLAP and statistics data cubes; dense data, with 100% or not much less of cells carrying data, such as satellite imagery.

Operations applied on such arrays can be studied by investigating image and signal processing, statistics, and linear algebra, to name a few. Finally, arrays regularly appear as "Big Data" with terabyte-sized single objects and petabyte archives, such as the holdings of Earth Observation (EO) data centers like the European Space Agency (ESA) and NASA archives.

Although arrays form an essential data structure in science and engineering and although this structure is well defined and known, database research has long neglected arrays, categorizing them as "unstructured data" to be stored as BLOBs. Consequently, no semantics and no operations can be offered by the database system, and hence users like large-scale data centers did not get any value from using databases for their array data. Still today, therefore, databases in science are mainly used for metadata while array and similar data are maintained by specially crafted data management tools with specialized service interfaces, but without fexible general-purpose query languages.

Only recently Array DBMSs have become a mainstream area of research. The pioneering system is rasdaman ("raster data manager") with its 24 years since its first publication and with a fully-fledged implementation used in operational installations since many years. Rasdaman is based on a minimal algebra on which query language, optimization, query evaluation, and storage layout is based. Among recent research approaches are SciQL, an array extension to the column-store MonetDB system, and SciDB, a standalone Array DBMS utilizing User-Defined Functions for providing array functionality.

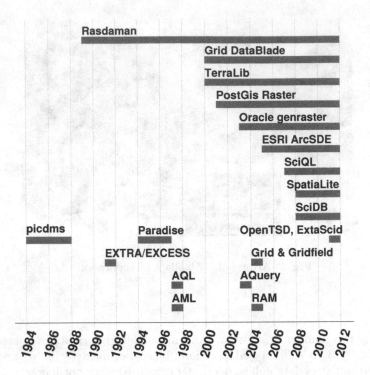

Fig. 2. A Brief History of Array DBMSs

In terms of data service standards, we can find arrays with ISO SQL:1998 and its successors and with the OGC Web Coverage Processing Service (WCPS) [7], a geo raster query language.

Given this relevance of array support in databases and with various, slightly varying systems emerging the quest for benchmarks arises. A comprehensive, well documented, and maintained benchmark can be of significant value to both deployers - like data centers - and database vendors, but also individual scientists and engineers. Further, it should allow to not only assess Array DBMSs as such, but also the large number of array supporting tools which are not using database technology, such as MatLab [19], R [23], and OPeNDAP [3].

In this paper we describe first concepts of a benchmark for large-scale, multi-dimensional array services. Currently, we are structuring the various facets of the possible and useful benchmarking tests. As a first result, we suggest a "suitability cube" framework in which all assessment aspects can be embedded. Under work is the refinement and breakdown of this concept.

In the next section we describe related work. Section 3 introduces the Suitability Cube. Practical examples are presented in Section 4; Section 5 concludes the contribution.

2   Related Work

Benchmarking of databases has been thoroughly addressed in the eighties and nineties, including periods of hot "benchmark wars" between vendors. Today, a set of generally accepted benchmarks is available for relational databases. Among the most popular are the SPEC [26] and TPC [10] database benchmarks.

There is no equivalent, though, for array databases. Given the only recently broadened interest of the database community there are no established benchmarks yet - actually, not even a commonly agreed conceptual data and query model.

Figure 2 gives a brief visual overview of the historical development. A notable precursor was PICDMS [9] which offered a conceptual model of a stack of same-resolution 2-D arrays with operators on them, although a generic array query language was not yet present, and no suitable architecture was indicated. Several publications emerged from relatively short-lived investigations. Most of today's systems, like PostGIS Raster [22] and rasdaman [18], add arrays as an additional attribute type, in sync with ISO SQL [14] which establishes arrays as a collection (i.e., column) type. A deviating approach is pursued by SciQL [16] and SciDB [24] where arrays are modeled similarly to tables, reusing much of standard SQL syntax albeit with a different semantics.

On commercial side, Oracle GeoRaster has to be mentioned, although - similar to PICDMS - it supports only 2-D arrays and lacks query support. ESRI ArcSDE has attempted to utilize databases for its 2-D rasters, but seems to not pursue development any further.

In terms of standards, we can find array support in two places. ISO SQL:1998 and its successors offer array support through an array collection type, although no array operators; a currently proposed new work item is aiming at closing this gap. The Open Geospatial Consortium (OGC) establishes and maintains Web service standards for geospatial intelligence. Arrays form a subcategory of so-called coverages [6], aka space-time varying phenomena. In 2008, OGC has adopted the Web Coverage Processing Service (WCPS) standard [7], a spatio-temporal geo raster query language, conceptually influenced by the rasdaman Array Algebra.

There have been several early attempts to benchmark geospatial databases [25,20], but these included e.g. a limited number of temporal queries or focused on domains like remote sensing exclusively so that evaluations outside their specific application domain were not feasible.

SS-DB has been proposed as a benchmark for science oriented databases[11]. By applying a space science use-case, it performs nine queries on astronomical array data. This case study is an important contribution towards understanding astrophysical workloads. However, the benchmark remains on application level and does not provide a thorough evaluation on model or algebra level. This benchmark has been run against SciDB [12] and MySQL [1]. It is available as open source, although similar results have not been reported yet by other groups.

For the geospatial domain, an analysis of relevant functionality has been pursued in [13]. Based on a broad survey of operations used in geo imaging, functionality has been classified and described by a uniform algebraic framework to allow for a systematic inspection. Representative array query examples further have been published for web mapping [4], genetic research [21], among others.

All these efforts are characterized by selecting particular use cases, without proof of covering the respective domain adequately. Further, there is no rigorous conceptual analysis of array queries which might characterize a concrete system's performance in its entirety. Therefore, the field currently is dominated by ad-hoc attempts. Our work aims at consolidating them into an Array Database benchmarking framework.

3 Benchmarking Arrays

3.1 Conceptual Array Modeling

Based on the common definition of an array as a function $a : X \to V$ from some d-dimensional Euclidean hypercube X into some value set V we naturally find some first query operation candidates:

- Changing the domain set X, often called subsetting; this can be differentiated into trimming (cropping the domain while retaining the number of dimensions) and slicing (extracting hyperslabs, thereby reducing dimensionality).
- Manipulating the value set V; this leads to a common set of unary and binary array operations, such as pixel-wise addition of images.
- Changing the array function itself, like establishing new mappings (examples include histograms and matrix multiplication).
- De-arraying functions, like aggregation.

When it comes to storing arrays, all systems uniformly perform partitioning - as practiced in image processing since long under the name *out-of-core processing* - into sub-arrays called chunks or tiles. Systems differ in the degree of variability. On one end there are static partitionings into square blocks where only the block size can be modified; on the high end are freely definable tiling schemes with and without overlapping, which forms an important tuning parameter. Also, these partitions naturally induce a tile streaming architecture which allows to keep only few parts of an array in server main memory during query evaluation, thereby achieving scalability in data volumes.

3.2 Benchmarking Dimensions

Based on the above outline of the concepts under test, we group features of an Array DBMS into several categories: Overall, we currently consider the following data categories as relevant for a benchmark:

- **Array Model Features:** Assess the expressive power of the data model:

- **Number of Dimensions:** This can be low-dimensional spatio-temporal data (1-D, 2-D, 3-D, 4-D, 5-D), medium-dimensional (6-D through 12-D), or high-dimensional (such as thousands of dimensions). Note that there is not a rigorous limit between boundaries, but we feel that the orders of magnitude separate good enough for focused testing.
- **Cell Type:** Array cells can contain single values, records of values (such as hyperspectral satellite imagery), as well as theoretically any other data structure. In practice, variable-length cell types like strings are avoided by all models inspected, due to the added complexity in storage management.

- **Array Data Properties:**
  - **Volume of Data:** Object sizes may range from a few kB for 1-D time-series over a few hundred MB for a satellite image up to PB size climate model output. Sizes of object sets can be massive as well – e.g., the European Space Agency (ESA) plans to have 10^{12} satellite images under their custody with their ngEO project.
  - **Sparsity of Data:** How does access and processing performance depend on sparsity, i.e., the percentage of non-null values within a data array. OLAP data, for example, have a density of typically 3% to 5% while satellite images often have a density of 100%.
  - **Storage Features:** What partitioning schemes does the Array DBMS support? Can partitions be compressed? Distributed?
- **Array Operations:** This encompasses questions like: what primitives are offered? Are operations executed natively or as UDFs? An open question is how to systematically scale query complexity for benchmarking.
  - **Isolated Position Relevance:** How does access to a large array depend on the size, shape, and position of the subsetting box?
  - **Coupled Position Relevance:** How does access to a large array vary when two subsettings are done in sequence, for different bounding boxes? What about non-trivial access patterns like in convolutions, statistical operations, Fourier Transforms, or simply mirroring an array?
  - **Processing Capabilities:** What array operations are offered? Formalizations like Array Algebra help to find comprehensive operations and operation combinations.
  - **Processing Implementation:** To what extent are array operations natively supported by the query engine, and where does it resort to UDFs? How efficient is the architecture, utilizing optimization, parallelization, etc.?
  - **Data Ingest and Update:** How fast can arrays be loaded? How fine-grain can updates to parts of arrays be applied?
- **Updates:** In view of the large size of single objects, it is not sufficient to only test creation, replacement, and deletion of whole objects. Updating an object typically will address selected areas within an array, which poses specific performance challenges.
- **Application Specific Features:** Geo imagery, for example, requires specific operations like orthorectification, coordinate transformation; statistical data require algebra operations like matrix multiplication and inversion.

4   Application Scenarios

Our research in array databases is based on both theoretical investigation, like finding a declarative, minimal Array Algebra [8], and extensive practical evaluations with users and in standardization bodies [5]. A number of domains in engineering and science have been investigated in close collaboration with large-scale data centers, including remote sensing, oceanography, geology, climate modeling, astrophysics, planetary science, computational fluid dynamics, genetics, and human brain imaging.

There is a diverse audience of users for these use cases. For the public at large, the database serves a large number of clients with typically a limited set of well defined queries wrapped in visual clients. Power users and researchers may use the database query language - possibly again with visual support - to conveniently wade through their raw data and run individual analyses.

Typically, the hardware to be used - like cloud, cluster and tape silos - is already present so the main question is not to determine the best hardware but to find the right data management and service tool for the given scenario. Here we hope to provide guidance with a reproducible benchmark.

5   Conclusion and Outlook

Arrays comprise an information category whose importance is just now being acknowledged by the database community at large. In science, engineering, business, social media, and statistics large arrays are of prime importance. With further systems emerging in addition to the pioneer Array DBMS, rasdaman, a standardized benchmark is useful for both system designers and data providers using such technology.

As part of ongoing activities towards a systematic benchmark we propose a first structuring of benchmark facets for a quantitative assessment of Array DBMSs. Aspects considered include conceptual data and query model capabilities, scalability in data volume, dimensionality, and query complexity, native query support vs UDFs, and application domain requirements. Therefore, we consider our work as a generalization of the application specific SS-DB performance comparison. In particular, experience from writing functional conformance test suites for geo raster services within OGC [2] has provided useful insights into test design and structuring.

Currently we are implementing the first slate of tests, focusing on storage access and array operations. Once a sufficient slate of tests is available, it is planned to run them against the available Array DBMSs and publish both benchmark code and results.

Acknowledgement. This research work is being supported by the European Community's Seventh Framework Programme (EU FP7) under grant agreement no. 283610 "European Scalable Earth Science Service Environment (Earth-Server)".

References

1. MySQL, http://www.mysql.com/
2. OGC Compliance Testing, http://www.opengeospatial.org/compliance
3. OpenNdap, http://www.openndap.org
4. Baumann, P., Jucovschi, C., Stancu-Mara, S.: Efficient map portrayal using a general-purpose query language (a case study). In: Bhowmick, S.S., Küng, J., Wagner, R. (eds.) DEXA 2009. LNCS, vol. 5690, pp. 153–163. Springer, Heidelberg (2009)
5. Baumann, P. (ed.): Web Coverage Processing Service (WCPS) Implementation Specification. No. 08-068r2, OGC, 1.0.0 edn. (2008)
6. Baumann, P.: Beyond rasters: introducing the new OGC web coverage service 2.0. In: Proceedings of the 18th SIGSPATIAL International Conference on Advances in Geographic Information Systems, GIS 2010, pp. 320–329. ACM, New York (2010), http://doi.acm.org/10.1145/1869790.1869835
7. Baumann, P.: The OGC web coverage processing service (WCPS) standard. Geoinformatica 14(4), 447–479 (2010), http://dx.doi.org/10.1007/s10707-009-0087-2
8. Baumann, P.: A database array algebra for spatio-temporal data and beyond. In: Pinter, R., Tsur, S. (eds.) NGITS 1999. LNCS, vol. 1649, pp. 76–93. Springer, Heidelberg (1999)
9. Chock, M., Cardenas, A.F., Klinger, A.: Database structure and manipulation capabilities of a picture database management system (picdms). IEEE Transactions on Pattern Analysis and Machine Intelligence 6(4), 484–492 (1984)
10. Council Transaction Processing Performance, TPC C Benchmark (2010), Standard Specification, http://www.tpc.org/tpcc/spec/tpcc_current.pdf
11. Cudre-Mauroux, P., Kimura, H., Lim, K.T., Rogers, J., Madden, S., Stonebraker, M., Zdonik, S., Brown, P.: SS-DB: A standard science DBMS benchmark (2010)
12. Cudre-Mauroux, P., Kimura, H., Lim, K.T., Rogers, J., Simakov, R., Soroush, E., Velikhov, P., Wang, D.L., Balazinska, M., Becla, J., DeWitt, D., Heath, B., Maier, D., Madden, S., Patel, J., Stonebraker, M., Zdonik, S.: A demonstration of SciDB: a science-oriented DBMS. Proc. VLDB Endow. 2(2), 1534–1537 (2009), http://dl.acm.org/citation.cfm?id=1687553.1687584
13. Garcia, A., Baumann, P.: Modeling fundamental geo-raster operations with array algebra. In: Proc. IEEE SSTDM, October 28-31, pp. 607–612 (2007)
14. ISO9075:1999: Information Technology-Database Language SQL. Standard No. ISO/IEC 9075:1999, International Organization for Standardization (ISO) (1999), (Available from American National Standards Institute, New York, NY 10036, (212) 642-4900)
15. Joachims, T.: Text categorization with support vector machines: Learning with many relevant features. In: Nédellec, C., Rouveirol, C. (eds.) ECML 1998. LNCS, vol. 1398, pp. 137–142. Springer, Heidelberg (1998), http://dx.doi.org/10.1007/BFb0026683
16. Kersten, M.L., Zhang, Y., Ivanova, M., Nes, N.: SciQL, a query language for science applications. In: Baumann, P., Howe, B., Orsborn, K., Stefanova, S. (eds.) EDBT/ICDT Array Databases Workshop, pp. 1–12. ACM (2011), http://dblp.uni-trier.de/db/conf/edbt/array2011.html#KerstenZIN11
17. Kimball, R., Caserta, J.: The data warehouse ETL toolkit. John Wiley & Sons (2004)

18. LSIS Research Group Jacobs University. The array database rasdaman, Rasdaman is available at http://www.rasdaman.org
19. MATLAB: version (R2013a). Natick, Massachusetts (2013)
20. Patel, J.M., Yu, J.B., Kabra, N., Tufte, K., Nag, B., Burger, J., Hall, N.E., Ramasamy, K., Lueder, R., Ellmann, C.J., Kupsch, J., Guo, S., DeWitt, D.J., Naughton, J.F.: Building a Scaleable Geo-Spatial DBMS: Technology, Implementation, and Evaluation. In: Peckham, J. (ed.) SIGMOD Conference, pp. 336–347. ACM Press (1997), http://doi.acm.org/10.1145/253260.253342
21. Pisarev, A., Poustelnikova, E., Samsonova, M., Baumann, P.: Mooshka: a system for the management of multidimensional gene expression data in situ. Inf. Syst. 28(4), 269–285 (2003), http://dx.doi.org/10.1016/S0306-4379(02)00074-1
22. PostGIS, http://www.postgis.org
23. R Core Team: R: A Language and Environment for Statistical Computing. R Foundation for Statistical Computing, Vienna, Austria (2013), http://www.R-project.org
24. Stonebraker, M., Brown, P., Poliakov, A., Raman, S.: The architecture of SciDB. In: Bayard Cushing, J., French, J., Bowers, S. (eds.) SSDBM 2011. LNCS, vol. 6809, pp. 1–16. Springer, Heidelberg (2011), http://dl.acm.org/citation.cfm?id=2032397.2032399
25. Stonebraker, M., Frew, J., Gardels, K., Meredith, J.: The Sequoia 2000 Benchmark. In: Buneman, P., Jajodia, S. (eds.) SIGMOD Conference, pp. 2–11. ACM Press (1993), http://doi.acm.org/10.1145/170035.170038
26. The Standard Performance Evaluation Corporation. SPEC Benchmark (2012), http://www.spec.org/benchmarks.html

Unleashing Semantics of Research Data

Florian Stegmaier[1], Christin Seifert[1], Roman Kern[2], Patrick Höfler[2],
Sebastian Bayerl[1], Michael Granitzer[1], Harald Kosch[1], Stefanie Lindstaedt[2],
Belgin Mutlu[2], Vedran Sabol[2], Kai Schlegel[1], and Stefan Zwicklbauer[1]

[1] University of Passau, Germany
[2] Know-Center, Graz, Austria

Abstract. Research depends to a large degree on the availability and
quality of primary research data, i.e., data generated through experiments and evaluations. While the Web in general and Linked Data in
particular provide a platform and the necessary technologies for sharing,
managing and utilizing research data, an ecosystem supporting those
tasks is still missing. The vision of the CODE project is the establish-
ment of a sophisticated ecosystem for Linked Data. Here, the extraction
of knowledge encapsulated in scientific research paper along with its public release as Linked Data serves as the major use case. Further, Visual
Analytics approaches empower end users to analyse, integrate and organize data. During these tasks, specific Big Data issues are present.

Keywords: Linked Data, Natural Language Processing, Data Warehousing, Big Data.

1 Introduction

Within the last ten years, the Web reinvented itself over and over, which led
from a more or less static and silo-based Web to an open Web of data, the so
called Semantic Web[1]. The main intention of the Semantic Web is to provide an
open-access, machine-readable and semantic description of content mediated by
ontologies. Following this, Linked Data [1] is the de-facto standard to publish
and interlink distributed data sets in the Web. At its core, Linked Data defines
a set of rules on how to expose data and leverages the combination of Semantic
Web best practices, e.g., RDF[2] and SKOS[3].

However, the Linked Data cloud is mostly restricted to academic purposes due
to unreliability of services and a lack of quality estimations of the accessible data.
The vision of the CODE project[4] is to improve this situation by the creation
of a web-based, commercially oriented ecosystem for the Linked Science cloud,
which is the part of the Linked Data cloud focusing in research data. This ecosystem offers a value-creation chain to increase the interaction between all peers,

---

[1] http://www.w3.org/standards/semanticweb/
[2] http://www.w3.org/RDF/
[3] http://www.w3.org/2004/02/skos/
[4] http://www.code-research.eu/

T. Rabl et al. (Eds.): WBDB 2012, LNCS 8163, pp. 103–112, 2014.

e.g., data vendors or analysts. The integration of a marketplace leads on the one hand to crowd-sourced data processing and on the other hand to sustainability. By the help of provenance data central steps in the data lifecycle, e.g., creation, consumption and processing, along corresponding peers can be monitored enabling data quality estimations. Reliability in terms of retrieval will be ensured by the creation of dynamic views over certain Linked Data endpoints. The portions of data made available through those views can be queried with data warehousing functionalities serving as entry point for visual analytics applications.

The motivation behind the CODE project originated from obstacles of daily research work. When working on a specific research topic, the related work analysis is an crucial step. Unfortunately, this has to be done in a manual and time consuming way due to the following facts: First, experimental results and observations are (mostly) locked in PDF documents, which are out of the box unstructured and not efficiently searchable. Second, there exist a large amount of conferences, workshops, etc. leading to an tremendous amount of published research data. Without doubt, the creation of a comprehensive overview over ongoing research activities is a cumbersome task. Moreover, these issues can lead to a complete wrong interpretation of the evolution of a research topic. Specifically for research on ad-hoc information retrieval, Armstrong et al. [2] discovered in an analysis of research papers issued within a decade, that no significant progress has been achieved.

In contrast to scientific events, ongoing benchmarking initiatives such as the Transaction Processing Council (TPC) exist. The main output of the TPC is the specification and maintenance of high-impact benchmarks for the database technology with members from Oracle, Microsoft, Sybase etc. Obviously, the industries are interested in running these benchmarks to show their competitive abilities. The results of those runs are published on the TPC website[5] as well as scientific workshops, e.g., Technology Conference on Performance Evaluation and Benchmarking (TPCTC)[6]. Unfortunately, it is currently very cumbersome to interact with the data to create comparisons or further visual analysis.

On the basis of this observations, the present issues could be improved by the use of the services established by the CODE project. Focusing on benchmarking initiatives, CODE technologies can be used to integrate the results of specific test runs, align them with extra information and therefore create an integrated TPC data warehouse to perform in depth analysis on the data, e.g., time series analysis.

This paper introduces the CODE project, along with its main processing steps. The main contributions are as follows:

- Data sources available in the research community will be described and the correlation to Big Data issues are given.
- The CODE project along with its main components is introduced with respect to the already defined Big Data processing pipeline.

---

[5] http://www.tpc.org/information/results.asp
[6] http://www.tpc.org/tpctc/

Table 1. Processable research data available in the CODE project

Type	Data Set Description	Data Characteristic
Research paper	PDF documents	Aggregated facts like tables, figures or textual patterns. Low volume, but high integration effort.
Primary research data	Evaluation data of research campaigns available in a spreadsheet like format (1. normal form) or via Web-APIs	Data generated by mostly automated means. Large volumes, low schema complexity.
Retrievable data	Linked Open Data endpoints	Semantically rich, interconnected data sets. Large volumes, hard to query (technically and from a usability point of view). Mostly background knowledge.
Embedded data	Microdata, Microformat, RDFa	Semantically rich, but distributed data. Less of interest.

The remainder of the paper is as follows: Section 2 highlights the data sources which can be processed by the CODE technologies. Here, an correlation to Big Data issues will be given. To get an understanding of the actual workflow, Section 3 proposes a processing pipeline, which is compliant to the overall definition of a *Big Data processing pipeline*. Finally, Section 4 concludes the paper and gives insights in the current achievements of the project.

2   Rediscovering Hidden Insights In Research

Research data is made available in various ways to the research society, e.g., stored in digital libraries or just linked to a specific website. Table 1 summarizes four data sources that are taken into account in the aforementioned usage scenario.

Research papers are a valuable source of state-of-the-art knowledge mostly stored in digital libraries reaching an amount of several Terabytes. Apart from the overall storage, the actual size of a single PDF document does not exceed a few Megabytes. The main task is to extract meaningful, encapsulated information such as facts, table of contents, figures and – most important – tables carrying the actual evaluation results. The present diversity of extracted data leads to a high integration effort for a unified storage. In contrast to that, primary research data is released in a more data centric form, such as table-based data. This kind of data is mostly issued by (periodic) evaluation campaigns or computing challenges. Famous examples are the CLEF initiative focusing on the promotion of research, innovation, and development of information retrieval systems. The outcome of such activities is thousands of raw data points stored in Excel sheets. Here, the volume of the data is most likely very large but defined by a specific schema with less complexity than PDF documents. Both data sources share an unstructured nature due to missing semantics on the schema and data level. To overcome this issue, the two remaining data sources of Table 1 are utilized. In this light, Linked Open Data endpoints serve as source for retrievable data, such as DBPedia[7] or PubMed[8]. On the one hand, these endpoints expose their data following the 5 star open data rule meaning the data is openly available, annotated with clear semantics and interconnected in the distributed Linked Open Data cloud. On the other hand, due to its distributed nature, efficient federated retrieval is a hard task. The last data source mentioned is embedded data meaning content of websites semantically annotated with microdata, microformat or RDFa. This information can be embedded table-based primary research data or auxiliary information, such as biographic data of a person.

As one can observe, there is a large amount of research data already available on the Web. The major drawback in this data landscape is the fact, that those are unconnected. Due to this fact, a comprehensive view is not possible, which leads to a loss of information. By the help of the CODE ecosystem, in particular by its data warehouse, this data gets connected and inference with respect to new knowledge is enabled.

Before considering the details of the knowledge extraction process, the correlation to the buzzword *Big Data* will be discussed. In todays research, the term Big Data[9] [3] is often used as a fuzzy concept without clear defined semantics. The following dimensions, the "3Vs", have to be mentioned when speaking of Big Data:

Volume is the most obvious characteristic for Big Data. Nearly every application domain produces an tremendous amount of data and is even increased by user interactions. This observation is also observable in terms of research data, when thinking of the amount of papers published with the correspoding monitored user interactions, such as citing.

Velocity makes it possible to state the production rate of the data. Huge data portions may be produced in real time in ongoing sensor systems, e.g., astronomy data, as an batch-like outcome of events, such as a conference or an evaluation campaign or single publications, such as white papers.

Variety takes the structure of the data itself into account. As already discussed, the data can be unstructured in silo-based PDF storage, semi-structured in Excel spreadsheets, or available in information retrieval systems.

Those three characteristics are commonly discussed by the community. It is clear, that Big Data at its core defines the data itself and the way it is processed and analyzed by corresponding pipelines. Linked Data on the other hand brings

---

[7] http://www.dbpedia.org/
[8] http://pubmed.bio2rdf.org/
[9] http://cra.org/ccc/docs/init/bigdatawhitepaper.pdf

in the techniques to semantically interlink and publish this heterogeneous portions of data. A recent white paper [4] issued by Mitchell and Wilson extend those Vs with respect to a data centric way:

Value of the data is the key to real interpretation and knowledge generation by answering the question which interaction steps of a processing chain made the data portions really "worthy".

The last characteristic can be directly aligned to the proposed approach. Here, crowd-sourced enabled data processing and analysis is combined with provenance chains to estimate the quality of the underlying data. By the help of Linked Data publishing techniques, the basis is given towards opening data silos for sophisticated interaction.

3 Big Data Pipeline Approach

When working with Big Data, Labrinidis and Jagadish [5] argue that "we lose track of the fact that there are multiple steps to the data analysis pipeline, whether the data are big or small". The Big Data processing pipeline proposed by CODE in terms of knowledge extraction of research data is illustrated in Figure 1.

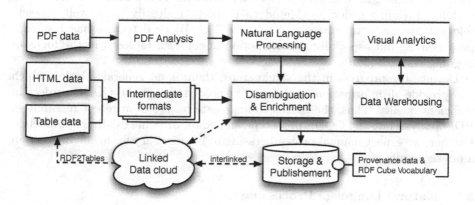

Fig. 1. Conceptual processing chain of knowledge creation and consumption

On the left hand side of Figure [5]the data sources introduced in Section 2 serve as an input for the conceptual processing chain. The data flow (continuous arrows) as well as dependencies (dashed arrows) are also plotted in the image. The central components are *PDF analysis, Natural Language Processing, Disambiguation & Enrichment, Data Warehousing* and *Visual Analytics* and will be discussed in the following.

3.1 PDF Analysis

Most of the research papers are stored in the PDF format. The quality of output of the PDF analysis thereby highly influences subsequent steps in the CODE processing chain. PDF is a page description language which allows low level control of the layout, but in this process the logical structure of the text is lost. For instance, text in multiple columns is often rendered across the columns, not adhering to the natural reading order. Especially tables are challenging because there is no annotation of logical tables defined in the PDF format. Still tables are assumed to contain lot of factual quantitative information. In general the challenges for PDF analysis can be summarised as:

- Text content extraction, extracting raw textual content (ignoring images and tables).
- Metadata extraction, e.g. extracting author names, titles, journal titles for scientific publications.
- Structure annotation, annotating document structure, e.g. for generating automatic table of contents.
- Block detection, detection of logical blocks like tables, abstracts.
- Table decomposition, extraction of table data according to its logical structure.

In recent years considerable research progress has been made with regard to these challenges. Text content extraction methods are able to extract text in human-reading order [6]. Metadata extraction already quite well extracts relevant metadata from scientific papers [7, 8]. Block detection has been approached [8], but especially the extraction of complex tables is in the focus of ongoing research [9, 10].

Despite the progress in the single steps, there is no general solution which can provide all information in the quality needed within the CODE project in sufficient quality. Thus, the task is to aggregate results from recent research on PDF analysis into the CODE prototype and adapt or refine existing approaches. Further, we expect manual post-processing to be necessary for achieving certain analysis results.

3.2 Natural Language Processing

Based upon the textual representation of a research article, the contained facts should be mined. Therefore techniques from the field of natural language processing are employed. As an initial step, named entities within the text are identified. Depending on the actual domain of the articles (biomedical domain, computational science, ...) the type of named entities varies.

Domain adaptation in the CODE project is foreseen to be transformed into a crowd-sourcing task. For example, in the computer science domain, where ontologies and annotated corpora are scarce, the users of the CODE platform themselves annotate the relevant concepts. Starting with the automatic detection of named entities, the relationship between those are identified in a second step. This way the textual content is analysed and domain dependant, factual information is extracted and stored for later retrieval.

3.3 Disambiguation and Enrichment

Entity disambiguation is the task of identifying a real world entity for a given entity mentioning. In presence of a semantic knowledge base, disambiguation is the process of linking an entity to the specific entity in the knowledge base.

Within the CODE project, entity disambiguation is applied to identify and link scientific knowledge artefacts mentioned in scientific papers. Subsequently background information from the Linked Science cloud can be presented to the user while reading or writing scientific papers.

The challenges regarding entity disambiguation within the CODE project are the following: (i) variance and specificity of scientific domains: not only do scientific papers cover a wide variety of topics but each domain very in-depth; (ii) synonyms in Linked Data repositories, and (iii) evolving knowledge: topic changes in scientific papers and in Linked Data endpoints.

Disambiguation using general purpose knowledge bases (mostly Wikipedia) has been widely covered in research, e.g. [11–13]. While approaches for specific knowledge bases exist, e.g. [14] for biomedical domain, the applicability of the approaches to a combination of general and specific knowledge bases and the resulting challenges (scalability, synonyms) has to be investigated within the CODE project.

After disambiguation, the gathered information for an entity can be extended by knowledge available in the Linked Data cloud. This extra information will be validated by user feedback and then integrated into the knowledge base. This process yields to an automatic and intelligent Linked Data endpoint facing the following research tasks: (i) integration and usage of provenance data, (ii) ranking and similarity estimations of Linked Data repositories or RDF instances, and (iii) quality of service parameter (e.g., response time). This process is often termed Linked Data Sailing. Currently, there exist frameworks to calculate similarity between Linked Data endpoints, e.g., SILK [15], and Linked Data traversal frameworks, e.g., Gremlin[10], which serves as a basis for further developments.

3.4 Storage and Publishing

The persistence layer of the CODE framework consists of a triple store, which has to offer certain abilities: (i) Linked Data compatible SPARQL endpoint and free text search capability, (ii) federated query execution, e.g., SPARQL 1.1 federated query[11], and (iii) caching strategies to ensure efficient retrieval. Those requirements are fulfilled by the Linked Media Framework [16], which has been selected for storage. For data modelling tasks, two W3C standardization efforts are in scope, which will be soon issued as official recommendations. The

---

[10] https://github.com/tinkerpop/gremlin/
[11] http://www.w3.org/TR/sparql11-federated-query/

PROV-O[12] ontology will be used to express and interchange provenance data. Further, the W3C proposes the RDF Cube Vocabulary[13] as foundation for data cubes, which are the foundation of data warehouses. Both vocabularies will be interconnected to ensure a sophisticated retrieval process.

3.5   Data Warehousing

As already mentioned, the basis for OLAP functionalities is the data cube. The data cube model is a collection of statistical data, called observations. All observations are defined by dimensions along with measures (covering the semantics) and attributes (qualify and interpret the observation). Well-known data warehousing retrieval functionalities would last from simple aggregation functions, such as *AVG*, up to high-level *roll up* or *drill down operators*. During retrieval the following functionality has to be ensured: (i) interconnection of RDF cubes, (ii) independence of dimensions, and (iii) high-level analytical retrieval in graph structures. Current research is dealing with the integration of RDF data into single data cubes [17, 18], but do not take an interconnection / federation into scope. Within the CODE framework, algorithms of relational data warehousing systems will be evaluated with respect to their applicability to graph structures. By the help of data cube interconnections complex analytical workflows can be created.

3.6   Visual Analytics

One important aspect of the CODE project is to make data available to end users in an easy-to-use way. This data might be already Linked Data as well as semantic data extracted from scientific PDFs. The goal is to build a web-based Visual Analytics interface for users who have no prior knowledge about semantic technologies. The main challenges regarding Visual Analytics in the scope of the CODE projects are:

- building an easy-to-use web-based interfaces for querying, filtering and exploring semantic data,
- developing semantic descriptions of Visual Analytics components to facilitate usage with semantic data, and
- building an easy-to-use web-based interfaces for creating visual analytic dashboards.

A query wizard is envisioned, with which users can search for relevant data, filter it according to their needs, and explore and incorporate related data. Once the relevant data is selected and presented to the user in tabular form, the Visualization Wizard helps them to generate charts based on the data in order to make it easier understandable, generate new insights, and communicate those insights in a visual way. One of the tools for visualizing the data will be MeisterLabs' web-based MindMeister mind mapping platform.

---

[12] http://www.w3.org/TR/prov-o/
[13] http://www.w3.org/TR/vocab-data-cube/

4   Conclusion

In this paper the challenges of the CODE project have been outlined. Further, the connection and the relevance to Big Data topics has been argued. In the current phase of the project, prototypes for certain issues of the introduced pipeline have been developed[14]. Within the second year of the project, those will be integrated into a single platform. Periodic evaluations will be conducted to ensure the required functionality and usability of the prototypes.

Acknowledgement. The presented work was developed within the CODE project funded by the EU Seventh Framework Programme, grant agreement number 296150. The Know-Center is funded within the Austrian COMET Program under the auspices of the Austrian Ministry of Transport, Innovation and Technology, the Austrian Ministry of Economics, Family and Youth and by the State of Styria. COMET is managed by the Austrian Research Promotion Agency FFG.

References

1. Bizer, C., Heath, T., Berners-Lee, T.: Linked data – the story so far. International Journal on Semantic Web and Information Systems 5(3), 1–22 (2009)
2. Armstrong, T.G., Moffat, A., Webber, W., Zobel, J.: Improvements that don't add up: ad-hoc retrieval results since 1998. In: Conference on Information and Knowledge Management, pp. 601–610 (2009)
3. Dumbill, E.: What is big data? An introduction to the big data landscape. O'Reilly Strata (January 11, 2012), http://strata.oreilly.com/2012/01/what-is-big-data.html
4. Mitchell, I., Wilson, M.: Linked Data - Connecting and exploiting Big Data. White Paper (March 2012), http://www.fujitsu.com/uk/Images/Linked-data-connecting-and-exploiting-big-data-(v1.0).pdf
5. Labrinidis, A., Jagadish, H.V.: Challenges and opportunities with big data. PVLDB 5(12), 2032–2033 (2012)
6. Hasan, I., Parapar, J., Barreiro, Á.: Improving the extraction of text in pdfs by simulating the human reading order. Journal of Universal Computer Science 18, 623–649 (2012), http://www.jucs.org/jucs_18_5/improving_the_extraction_of
7. Granitzer, M., Hristakeva, M., Knight, R., Jack, K., Kern, R.: A comparison of layout based bibliographic metadata extraction techniques. In: Proceedings of the 2nd International Conference on Web Intelligence, Mining and Semantics, WIMS 2012, pp. 19:1–19:8. ACM, New York (2012)
8. Kern, R., Jack, K., Hristakeva, M.: TeamBeam - Meta-Data Extraction from Scientific Literature. D-Lib Magazine 18 (July 2012)
9. Fang, J., Gao, L., Bai, K., Qiu, R., Tao, X., Tang, Z.: A table detection method for multipage pdf documents via visual seperators and tabular structures. In: 2011 International Conference on Document Analysis and Recognition (ICDAR), pp. 779–783 (September 2011)

---

[14] http://www.code-research.eu/results

10. Liu, Y., Bai, K., Gao, L.: An efficient pre-processing method to identify logical components from pdf documents. In: Huang, J.Z., Cao, L., Srivastava, J. (eds.) PAKDD 2011, Part I. LNCS (LNAI), vol. 6634, pp. 500–511. Springer, Heidelberg (2011)

11. Kataria, S.S., Kumar, K.S., Rastogi, R.R., Sen, P., Sengamedu, S.H.: Entity disambiguation with hierarchical topic models. In: Proceedings of the 17th ACM SIGKDD International Conference on Knowledge Discovery and Data Mining, KDD 2011, pp. 1037–1045. ACM, New York (2011)

12. Fader, A., Soderl, S., Etzioni, O.: Scaling wikipediabased named entity disambiguation to arbitrary web text. In: Proc. of WikiAI (2009)

13. Dredze, M., McNamee, P., Rao, D., Gerber, A., Finin, T.: Entity disambiguation for knowledge base population. In: Proceedings of the 23rd International Conference on Computational Linguistics, COLING 2010, Stroudsburg, PA, USA, pp. 277–285. Association for Computational Linguistics (2010)

14. Rebholz-Schuhmann, D., Kirsch, H., Gaudan, S., Arregui, M., Nenadic, G.: Annotation and disambiguation of semantic types in biomedical text: a cascaded approach to named entity recognition. In: Proceedings of the EACL Workshop on Multi-Dimensional Markup in NLP, Trente, Italy (2006)

15. Volz, J., Bizer, C., Gaedke, M., Kobilarov, G.: Discovering and maintaining links on the web of data. In: Bernstein, A., Karger, D.R., Heath, T., Feigenbaum, L., Maynard, D., Motta, E., Thirunarayan, K. (eds.) ISWC 2009. LNCS, vol. 5823, pp. 650–665. Springer, Heidelberg (2009)

16. Kurz, T., Schaffert, S., Bürger, T.: LMF – a framework for linked media. In: Proceedings of the Workshop on Multimedia on the Web Collocated to i-KNOW/ i-SEMANTICS, pp. 1–4 (September 2011)

17. Kämpgen, B., Harth, A.: Transforming statistical linked data for use in olap systems. In: Proceedings of the 7th International Conference on Semantic Systems, I-Semantics 2011, New York, NY, USA, pp. 33–40. ACM (2011)

18. Zhao, P., Li, X., Xin, D., Han, J.: Graph cube: on warehousing and olap multidimensional networks. In: Proceedings of the International Conference on Management of Data, pp. 853–864 (2011)

Generating Large-Scale Heterogeneous Graphs for Benchmarking

Amarnath Gupta

San Diego Supercomputer Center
Univ. of California San Diego
La Jolla, CA 92093, USA

Abstract. Graphs have emerged as an important genre of data that are found in a wide class of applications. The most dominant benchmark for graph data today is Graph 500 that generates a Stochastic Kronecker graph of various sizes, and reports the time to perform a breadth-first search. Apache Giraph uses Pagerank computation as an algorithmic benchmark for large graphs, but does not provide the mechanism to generate graph data. Other forms of graph benchmarks have been developed by smaller communities and are not known widely. However, most benchmarking data for graphs are derived from a single structure generation model, and therefore does not capture the variability of structure and content. To this end, we propose *heterogeneous graphs*, a mixed model graph structure that combines several existing generation techniques into a single benchmark. It is a hybrid that constructs edge-labeled multi-graphs with multiple components, which can be hierarchical, power-law graphs, community-forming graphs, and a new class of graphs formed by motif composition. The user can use a simple set of 4 parameters to specify the graph, but has the option to use several more parameters to have a finer control of the hybrid structure. We define the generation process for heterogeneous graphs and propose an initial set of query operations against the generated data.

Keywords: heterogeneous graph, benchmarking, power law, community structure, data generation.

1 Introduction

The unprecedented growth of the social media industry in the past few years have cast a spotlight on the importance and usefulness of graph data. In December 2012, Facebook reported 1.06 billion monthly active users and 618 million daily active users, while in January 2013, LinkedIn reported more than 200 million acquired users. With these explosive numbers, the industry is finding new ways to utilize and productize graph-based analysis. Products for tasks like finding communities with a specific demographic and interest profile, finding objects (e.g., pictures, places, products, ...) that like-minded people use, finding highly networked people with a certain expertise, are beginning to emerge. At the same time, the market is quickly finding new ways to utilize these products;

T. Rabl et al. (Eds.): WBDB 2012, LNCS 8163, pp. 113–128, 2014.
© Springer-Verlag Berlin Heidelberg 2014

for example, marketing companies are beginning to exploit the knowledge of discovered communities to identify their targets. Taking a step back, one can see that despite the recent surge of popularity and interest, graph-structured data had always been used in academic research and niche product markets. Many complex graph data manipulation and analysis algorithms have been developed by the Computer Science community. Perhaps more interestingly, application communities have done extensive research in using graph data for their specific problems. In Biomedical Sciences, researchers study patterns of connectivity and behavior for interconnected biological objects that evolve with time; in Data Mining and Knowledge Discovery, researchers study call logs and discover "hidden" patterns in mobile communication graphs; in Linked Graph communities Systems Analysts develop navigation and exploration techniques over a wide range of connected data sets; Social Science researchers study network influences and create metrics for scoring individuals or groups based on their "dynamics" (i.e., activity patterns) on any social networks. These groups have created many algorithms for simulating, storing, partitioning, navigating, searching, indexing, summarizing and analyzing various forms of graphs. However, these algorithms have mostly been used for the sizes and variants of graph data they needed for their own purposes. Now, with the advent of new use cases and market demands for huge and rapidly increasing graph sizes, one needs to determine what kinds of algorithms are required, and what kind of operating infrastructure these algorithms should run on in practice to achieve the desired performance at the scale needed. This motivates us to consider graph data benchmarking as an important genre within the larger context of Big Data benchmarking. A graph data benchmarks will contributed to the fair and standardized assessment of performance across different algorithm providers, different system versions from the same provider and across different architectures.

Challenges. Bhandarkar[1] argues that big data systems are characterized by their flexibility in processing diverse data genres, including graphs, geo-locations and text, using a variety of methods. Because of the many sources and methods of analyzing Big Data, *a single benchmark that characterizes all use-cases could not exist*. We hold that the same is true even within the genre of graph data, and developing a single benchmark that represents the structure and processing needs of all graph data is impossible to create. For example, social networks are structured differently biological interaction networks, and have different evolution processes when viewed over time. Although they share some common forms of query and analysis (e.g., finding "influential nodes") are common to both forms of graphs, significant differences exist between them. For example, social networks are characterized by small clique-ish groups, while biological networks are often used for subgraph pattern matching (motif discovery).

In this paper, we consider the problem of large-scale graph data generation for benchmarking. The contributions of this work are as follows.

---

[1] http://reflectionsblog.emc.com/2013/02/industry-standard-benchmarks-for-big-data-platforms.html

- Based on a number of application scenarios, we develop a set of general-purpose design guidelines for graph data generators.
- Using these guidelines, we construct a specific application case to create a graph generator.
- We present the specification and a reference implementation of the generator for directed graphs with labeled edges with constraints.

2   Related Work

The most well-known graph benchmark till date is Graph 500 (www.graph500.org), which currently intends to provide benchmark data sets for three application kernels: concurrent search, optimization (single source shortest path), and edge-oriented (maximal independent set) in the context of five graph-related application areas: Cybersecurity, Medical Informatics, Data Enrichment, Social Networks, and Symbolic Networks. In this section, we describe the Graph 500 data generator. We also present the data generators for some more specialized benchmarks that target different audiences.

Graph 500. For Graph 500, the user provides two parameters called SCALE and EDGEFACTOR. The system uses this to create a graph G with $N = 2^{SCALE}$ nodes and $E = $ EDGEFACTOR $* N$ edges. The goal is to construct the graph using a graph generation technique called Kronecker generator similar to the Recursive MATrix (R-MAT) scale-free graph generation algorithm [1]. The R-MAT generator uses an adjacency matrix data structure. It recursively sub-divides the adjacency matrix of the graph into four equal-sized partitions and distributes edges within these partitions with unequal probabilities. Initially, the adjacency matrix is empty, and edges are added one at a time. Each edge chooses one of the four partitions with probabilities A, B, C, and D, respectively. The graph generator creates a small number of multiple edges between two vertices as well as self-loops. It also generates the data graph with high degrees of locality. However, according to the Graph 500 standard the vertex numbers must be randomly permuted, and then the edge tuples must randomly shuffled to remove the high degree of locality. This last requirement ensures that no benchmark algorithm can exploit the locality to their advantage. The primary benchmarking algorithm test performed on the data is a breadth-fisrt traversal that starts from an arbitrary node and constructs the BFS tree of its traversal. More recently, [2] has developed a stoachastic generalization of the R-MAT method that has been empirically observed to have interesting real-network-like properties.

The LFR Benchmark. Based on the "mixing pattern" model of [3], Lanci-chinetti, Fortnato and Radicchi (hence LFR) [4] develop a class of benchmark graphs that model graphs whose nodes participate in internal community structures. The benchmark models real-world networks (e.g., social networks) containing communities of different sizes. To realize this, the algorithm assumes that both the degree and the community size distributions are power laws, with exponents β and γ, respectively. Each node is given a degree taken from a power law distribution with exponent γ. The extremes of the distribution k_{min} and

k_{max} are chosen such that the average degree is k. Each node shares a fraction $(1 - \mu)$ of its links with the other nodes of its community and a fraction μ with the other nodes of the network, where μ is called *mixing parameter*. The sizes of the communities are taken from a power law distribution with exponent β, such that the sum of all sizes equals the number N, i.e., the number of nodes of the graph. The generation process starts with an empty graph and incrementally fills in the adjacency matrix by obeying the constraints above. The target of this benchmark is to evaluate algorithms that attempt to find community structures in a network – utilized in solving "people finding" tasks through citation, co-participation, and professional networks. We note that unlike the Graph 500 case, these graphs are supposed to have local internal structures, and therefore serve a different purpose as a benchmark for a different (and growing) segment of the graph-data industry.

The S3G2 Method. The previous two methods generated graphs whose nodes and edges were not labeled. Thus, they cannot be used to generate database-like graphs such as RDF data (or any other data that can be viewed as RDF). [5] proposes a method called Scalable Structure-correlated Social Graph Generator (S3G2) that addresses the problem of generating scalable random graphs with value and structure correlations. In this model, the graph generator produces new nodes with property values, and edges between these nodes and existing nodes. In S3G2 graph, a node belongs to one of the object classes or is a literal. A labeled edge contains two nodes and an edge property in which one node belongs to an object class and the other node is a literal or an object. The edge property is a literal property or a relation property, respectively. One node can have many edges with the same edge property and there is no edge connecting two literal nodes. In an edge, the end node is considered as the property value of the start node. The generation algorithm uses a set of *correlation rules* to construct legitimate edges. In doing so, it uses a probability model to choose a certain value from a dictionary, or the probability to connect two nodes with an edge are thus influenced according to these correlation rules, by existing data values. For instance, the birth location of a person influences probability distribution of the first name and school dictionaries. As another example, the probability to create a friendship edge is influenced by agreement on birth-year and school properties of two person nodes. The benchmark uses mapreduce style algorithms to generate social network-like graphs that network analysis algorithms can use. There are other well-known benchmarking standards, like the LUBM [6] and the BSBM [7] benchmarks for RDF and OWL data, that might be considered as graph data depending on application and implementation. For example, some implementations can treat them as Description Logic systems that do not have any explicit graph operations, while others, like the linked-data systems that emphasize on traversal based access to graph nodes using a query language like SPARQL, do model RDF data as graphs. This makes them relevant yet somewhat out of our scope for this paper. The primary observation we would like to make from these use cases is that with graphs, it is insufficient to construct a single application-level benchmark scenario like TPC-C that is representative enough for most applications.

Next, we make a case for developing a data generator for *heterogeneous graphs*, that we believe are more pervasive than we generally acknowledge, and represent a new class of applications that are underrepresented in literature. We motivate the case with an application scenario.

3 An Application Scenario

Our application is modeled after a drug discovery scenario, where a research organization maintains both its private data as well as publicly available data it has gathered from different web sites. We deliberately choose a scientific rather than a business scenario for our benchmark because we believe that scientific applications are a rich, vital yet underserved territory for the benchmarking community and expect that efforts like this will help foster the development of interesting algorithm and system design in the future. The benchmark is based upon the following considerations.

- The data is structured as a combination of N overlapping named graphs $G_1 \ldots G_N$, where the overlap is accomplished by node sharing. The nodes of the graphs represent instances of biological entities like genes, proteins, parts of the human body, pharmacological compounds, and so forth, while the labeled edges represent (a) attributes that have scalar values (of type int, string, float ...), and (b) binary relationships between entity pairs. *They overlap because each named component graph is independently produced by different user groups who populate different properties of the same entities.*
- A subset of the named graphs $G_1 \ldots G_k$ are hierarchical, i.e., they are structured as trees or DAGs. Physically, they represent class hierarchies and partonomy hierarchies among entities. For example a drug classification system is a hierarchical graph. Similarly, a citation network is also a hierarchical graph (DAG). Each hierachical graph uses a single hierarchy-forming relationship (HFR) (i.e., only one edge label is used in each hierarchical graph). For the drug classification case, the HFR is subclassOf, while in the citation case, the HFR is cites.
- The remaining $N-k$ graphs are multigraphs (i.e., there can be multiple edges between two nodes). For example, the research organization would often download biological interaction graphs from the National Center for Biomedical Informatics (NCBI). This network contains information about how biomolecules interact with each other, and may have recors for two molecules A and B that satisfy the relations (A physically-interacts-with B) and (A positively-regulates B). However, we would like to ensure that there is no redundant content in the graph. So we impose the constraint that there cannot be two different edges with identical labels between two nodes for the same graph. Note however that since we assume the component graphs to be independently created, this does not preclude two different graphs from creating two edges of the same label on the same two nodes.

– The multigraphs differ in terms of their network connectivity properties.
  • Some component graphs (biological networks) obey the power-law more
    strictly than others (human social networks e.g., science groups)
  • some graphs have a larger skew in the distribu-
    tion of edge labels (more specialized properties like
    (*geneA* is-allelic-variant-of *geneB*) are less abundant than
    generic properties like (*proteinA* is-target-of *drugB*))
  • some graphs (physical interactions among molecules) are denser (i.e.,
    have a higher node to edge ratio) than others (citation networks)
  • some graphs may optionally have additional constraints regarding sub-
    graph patterns, containing patterns that ought to appear, and patterns
    that are prohibited. For example,

These characteristics illustrate why generating benchmark data for heterogeneous graphs cannot be treated the same way as relational or warehouse data
benchmarks where the variability in data patterns and its impact on access operations are less pronounced. It also justifies why generation of benchmark data
for heterogenous graphs should not use a fixed schema pattern that most TPC
standards use, but should rather be based on a number of graph parameter
characteristics the above examples allude to.

In the next section we present our benchmark.

4   GDB-H: The Heterogeneous Graph Data Benchmark

The Setting. The drug discovery lab modeled in the benchmark is interested in 11 entity categories: genes, proteins, diseases, anatomy (includes gross
anatomical parts, tissues, cells, and subcellular structures), phenotypes (i.e., observed characteristics), drugs (i.e., pharmacological substances that may have
multiple brand names), interactions, pathways (named interaction graphs over
genes/proteins), species, experiment types, and publications. In Table 1, these
entity types are associated with the following sets of attributes expressed in an
object-relational style.

In this schema, multivalued attributes are designated as a set of a type, and
object references are designated as a set of object identifiers. In addition to
these categories, the benchmark assumes a set of objectClassNames for each
category. For the sake of convenience we will use generic names for these classes.
For example, class names for interactions will be called interactionClass-1,
interactionClass-2, ... and so forth.

Not included in the schema is a list of 3000 binary relationships which we call
$R_1 \ldots R_{3000}$ here, including HFRs that hold between pairs of instances of these
entities. Table 2 shows a few of these relationships.

The Benchmark Data. The GDB-H benchmark allows a user to specify a
small number of mandatory parameters and a larger set of optional paramters.
The mandatory parameters are:

(a) GRAPHNUM - total number of component graphs with a minimum value of
    8, and maximum value of 100.

Table 1. The entity types and their attributes used in GDB-H

Entity Type	Attributes
genes	id:int, symbol:char[4], name:char[32], organism:char[32], chromosome:[2], startPosition:long, endPosition:long
proteins	id:int, symbol:char[4], name:char[32], molecularWeight:float, producingGene:genes.id, function:set(char[32])
diseases	id:int, name:char[32], broadType:char[32], description:char[1024], cause:char[1024], signSymptoms:char[1024], affectedPopulation:string
anatomy	id:int, name:char[32], isGrossAnatomicalObject:bool, isTissue:bool, isCell:bool, isSubCellular:bool
phenotype	id:int, name:[32], description:[256], affectedAnatomicalPart:set(anatomy.id)
drugs	id:int, substanceName:char[128], brandName:char[128], targetDisease:set(disease.id), targetProtein:proteins.id, targetBiologicalProcess:set(char[128])
interactions	id:int, sourceMolecule:char[256], targetMolecule:char[256], interactionType:char[128], referenceCitation:set(publications.id)
pathways	id:int, name:char[128], description:char[512], pathwayGraph:set(interactions.id)
species	id:int, commonName:char[128], scientificName:char[256]
experimentType	id:int, name:char[128], description:char[256]
publications	id:int, title:char[256], authors:set(char[1024]), journal:char[512], date:dateTime

Table 2. A representative sample of the relationships used by the benchmark

Relationship	Domain	Range	HFR?	Explanation
instanceOf	entityInstance	ObjectClassName	no	example: g1 is an instance of genes
subclassOf	objectClassName	ObjectClassName	yes	example: drug is a subclass of molecules
partOf	anatomy	anatomy	yes	example: finger is a part of hand
cites	publication	publication	yes	
follows	interaction	interaction	yes	interaction i1 occurs after interaction i2
is-target-of	protein	drug	no	example: drug d1 works by affecting protein p1
is-inside	anatomy	anatomy	yes	example: brain is inside the skull

(b) NODES - total number of nodes, with a minimum of 100,000, and a maximum of 100,000,000.

(c) NODE-TYPES - total number of node types, a value between 3 and 11, where 11 will consider all entity types shown in Table 2.

(d) EDGE-LABELS - total number of distinct edge labels, a value between 30 and 3000.

Note that we do not use the edge-factor parameter used by Graph 500, primarily because the above parameters are sufficient to create a heterogeneous data graph using the method described in the next section. If all values are chosen at the minimal level, the total data size will be approximately 10^{10} bytes, the same as the toy size of the Graph 500 benchmark, and choosing the maximum value will correspond to 10^{14} bytes, the large size of the Graph 500 benchmark. However, a more sophisticated user will use additional parameters to control the nature of the graphs produced. These parameters are:

(i) TYPE-RATIOS - a set of (property, proportion) pairs where properties are hierarchical, power-law [8], community [9], motif composition (see next section), and the proportions are non-zero numbers that denote the fraction of the component graphs that should have these properties. The proportion values must add up to 1. Note that in an extreme (and undesirable) case where TYPERATIOS= (hierarchical, 1.0), the system generates a number of different classifications on the same set of terminal nodes. In the default case, this proportions are determined by the system, and favors the power-law and community-structured graphs. As a further option one may choose the desired power values of the power-law graphs, however, the system will adjust these values (and drop them if the number of component graphs does not match) to fit all constraints on the specification.

(ii) NODE-DISTRIBUTION - a set of qualitative statements that characterize the relative sizes of the component graphs. It is specified as a set of (size-property, proportion) pairs where the size-properties are small, medium, large, very large, huge and the proportions, as before, denote the fraction of the component graphs that should have these properties. In practice, this specification will be taken as a guideline for node allocation to the component graphs. In the default case, the minimal setting will produce a limited non-uniformity in the node sizes of the distribution. With increasing node count, the full graph will be more diverse in node distribution.

(iii) EDGE-DENSITY - as with NODE-DISTRIBUTION, this property sets the guideline for the expected relative densities of the edges. The edge density is measured as $D = 2|E|/|V|(|V| - 1)$, where $|V|, |E|$ are the node and edge counts respectively; for the specification edge density is grouped into the classes very sparse, sparse, medium, dense, very dense.

(iv) OVERLAP - this parameter specifies the degree of node overlap that the different component graphs of the total graph. The parameter is specified as (degree of overlap, propotion) pairs, where the degree of overlap is specified as none, light, medium, heavy, full. However, there are some restrictions to the manner in which the specification is applied. A hierachical graph can never be `full` or `none`, thus making a declaration like OVERLAP= `(full, 1.0)` will be considered an error, unless the proportion of hierarchical components is set to 0.

(v) LABEL-TO-EDGE - the label-to-edge ratio $R = |L|/|E|$ of a graph, where L is the number of distinct labels and $|E|$ measures the degree of edge coloring in a labeled multigraph. For hierarchical graphs $|L| = 1$ by definition, making $1/|E|$ the lower bound for R. As in the case of the previous measures, the user provides a qualitative specification of this ratio using the categories `low`, `medium low`, `medium`, `medium high`, `high`, and provides the proportion of graph components for each category. Thanks to the size dependent nature of this ratio, the `low` category is interpreted to be a small multiple of $1/|E|$.

(vi) CONSTRAINTS - a set of first order predicates that must hold for a user-specified graph. For the purposes of this benchmark, the constraints may only be based on edge patterns, node and edge properties, or cardinality of nodes and edges. We present a few illustrative examples:

```
- if ((node1 subClassOf node2), (node2 subClassOf node3))
  disallow ((node1 subClassOf node3))
- if (node1 is-target-of node2)
  disallow ((node1 is-target-of node2), (node1 is-target-of node3), (node1 not node3))
- if ()
  disallow((indegree(node1)>10), (node1 :R node2), (indegree(node2)>10))
- if (node1 instanceOf genes), (node2 instanceOf proteins), (node1.symbol
  = node.symbol)) mustOccur ((node2 derived-from node1))
- if ((node2 subClassOf node1), (count(node2) > 20))
  mustOccur((node2 IN componentGraph:G), (nodeCount(G) > 200))
```

Here all constraint expressions are inspired by the SPARQL syntax, and use edge triples, and `node1, node2` are treated as node variables. The first constraint prohibits the matrialization of the transitive closure of the HFR `subClassOf`. The second constraint uses a simple form of negation to disallow the incidence of two distinct edges of a specified label to the same node. The third constraint has an empty `if` clause, and disallows two connected nodes to each have a high indegree. Note the use of the builtin function `indegree`, as well as the use of the unbound edge label `:R` in the constraint. The fourth constraint shows the use of `mustOccur` that asserts a consequent edge pattern given an antecedent edge pattern. Here, we also show the use of the `node.property` construct, which are used for the static attributes of the entity types from Table 1. Finally, shows the use of two aggregate functions `count(node)` (resp. `count(edge)`, and `nodeCount(graph)` (resp. `edgeCount(graph)` in a constraint. Further, node (resp. edge) IN graph is a membership function that relates a node to the component graph it belongs to. Since the constraints are not specific to any component graph, the benchmark applies them to the entire graph.

We believe that making these parameters optional does not place significant burden on the users who would like to run their algorithmic benchmark on the default graph, and yet having them available provides a rich set of parameters for more exploratory users who currently have no flexible mechanism of generating complex heterogeneous test graphs and run queries against them.

5 Generating GDB-H Graphs

The central task is GDB-H benchmark generation is to create random labeled graphs that are hierarchical, power-law, community-structured or purely random, over a set of overlapping nodes whose entity types come from Table 1, and construction criteria follow the user directives specified in the last section. We achieve this by combining a number of existing graph generation models for these 4 graph categories, duly customized to suit our benchmarking requirements. We first describe our base-level models for these categories.

Hierarchical. Our base model for hierarchical graphs derives from [10]. This model holds that a DAG is acyclic because there is an underlying ordering amongst the nodes (e.g., due to time in citation graph). We use the case from [10] where the edge probabilities are independent of each other. For this model, one generates a Poisson distributed random number m with mean equal to the desired expected number of edges (which we compute by , then distribute those edges at random over the graph in proportion to P_{ij}, which is the probability that a certain edge emanating from node j will connect to a node $i, (j > i)$. For GDB-H, we first need to estimate m. To do this we assume that the average in-degree of a node is 5, and we generate the indegree k_i^{in} for node i based on a Gausssian distribution with $\sigma = 2$, rounded off to the closest integer. Then $m = \sum_{i=1}^n k_i^{in}$. Now we need to add more semantics to this model. For example, if the hierarchy is to represent a classfication (i.e., a subclass of relationship), there is no total order on the node set; hence we impose an arbitrary order amongst the nodes. If the hierarchy is to represent a partonomy, then we restrict the model such that an edge does not go from some node j to a node $i - k$ bypassing node i. On the other hand, if it is a citation graph between pairs of publication entities, the nodes are ordered by the actual date stamps of the randomly generated instances of publication. The model can generate a small number of duplicate edges between the same two nodes – we simply eliminate these edges.

Power-Law. Power-law, i.e., the assertion that the number of nodes of a graph, y, of a given degree x is proportional to x^β for some constant $\beta > 0$, is a very popular model for natural graphs. It has been empirically shown that many biological interaction graphs approximately show a power-law distribution. Although the theory has been significantly criticised in recent years [11], the power-law nonetheless remains a reasonable rough model of biological interactions. We produce power-law graphs by following the generative model in [8]. By this model, the maximum degree of the node is $e^{\alpha/\beta}$ where α is the logarithm of the size of the graph. We keep $2 < \beta < 3.47875$. For this range the number of nodes $n \approx \zeta(\beta)e^\alpha$ where $\zeta(x)$ is the Riemann zeta function. Since n is given for us, we can approximately compute the value of α, and consequently the number of edges $m = 0.5\zeta(beta - 1)e^\alpha$. Keeping β between 2 and 3.47875 ensures that there is a unique giant component of the graph, but also there is very likely a second large component, and the whole graph is most likely connected. Other ranges of β, particularly $1 < \beta < 2$, has been found useful in some realistic models of gene interaction systems [12], but we do not use this range because it might produce multiple unconnected components of the graph, which is not our goal. To customize this model for our requirements, we use the user-defined parameters NODE-DISTRIBUTION and EDGE-DENSITY to determine the value of n and β respectively (in practice, we use stored values of these parameters from pre-run simulations).

Community Model. It has been shown that unlike the pure power-law graph, many real graphs show the formation of community structures, in which if nodes A and B are related and C is related to one of them, it is very likely that C is also related to the other, thus forming an ABC community. [9,13] formalize this model and propose a generative model for graphs with community structures. They measure the global as clustering coefficient c as the ratio: $c = no.$ *of closed wedges/no. of wedges* and the clustering coefficient per degree as the fraction:

$$c_d = \frac{no.\ of\ closed\ wedges\ centered\ at\ a\ node\ of\ degree\ d}{no.\ of\ wedges\ centered\ at\ a\ node\ of\ degree\ d}$$

If n_d is the number of nodes of degree d, the total number of nodes $n = \sum_d n_d$ and the number of edges $m = 0.5 \sum_d d.n_d$. In our case, we are given n, so we need to create the degree distribution n_d. Fortunately, it is shown [9] that n_d has a log normal distribution, i.e., $n_d = k.exp(-(\frac{log\ d}{\alpha})^\beta)$, where it has been experimentally determined that $1.85 < \alpha < 2.2$ and $1.75 < \beta < 2.1$. The factor k is set based on NODE-DISTRIBUTION. It has also been found that the mean value of c_d is $\bar{c}_d = c_{max}.exp(-(d-1)^p)$ where c_{max}, the maximum clustering coefficient and p are set based on EDGE-DENSITY using precmputed simulation results. Then the distribution c_d is given by the normal distribution $\mathcal{N}(\bar{c}_d, min(0.01, 0.5\bar{c}_d))$. The construction of the graph follows the BTER generation technique described in [9]. As before, we eliminate any duplicate edges generated in the process.

Motif Composition. In contrast to the previous approaches, which can be viewed as construction of graphs with global structural constraints, motif composition is a "bottom-up" construction process. It uses a library of motifs, i.e., commnly occurring local graph patterns and composes them to produce a larger graph. The motivation for providing this construction option is that for several kinds of networks, it is very difficult to have overarching models that reflect the local patterns faithfully. In our application domain, one such class is biological pathways which could be regulatory, signal transduction or metabolic by functionality. Although we do not have any empirical proof at this point, we conjecture that the same requirement of matching local patterns holds for other domains as well. We approach the problem by constructing a predefined motif library. Figure 1 illustrates some of the patterns observed in metabolic pathways [14]. Each motif has open-ended placeholders where other motifs can be fit. The motif composition graph is generated by computing the total number of motifs that should be used based on the expected size of the graph and then selecting a random selection of motifs such that the total number of open ends is minimal. This means that the i-th motif's placement in the graph is conditional upon the placement of the open slots left behind by the previous $(i-1)$ motifs. For this case, the EDGE-DENSITY parameter is ignored. Also, notice the cyclic

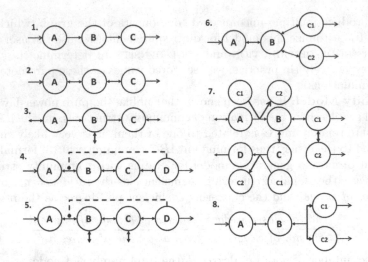

Fig. 1. A set of motifs for metabolic network, adapted from [14]

relationship in the patterns. Our generation process ensures that *while cycles may exist in the data, no cycles exist for the same edge label in a motif-composed graph.*

Graph Generation. Based on the above considerations for generating the component graphs of various types, the overall graph construction process is as follows.

1. Based on GRAPHNUM, NODES and NODE-DISTRIBUTION we allocate the number of graphs per per graph type and the number of nodes per graph such that in the default case, node counts are ordered as hierarchies < motif composition graphs < community graphs ≈ power-law graphs. This can however be altered based on the TYPE-RATIO specification.

2. For each node set allocated to power-law and community-structured graphs
   (a) We allocate a subset of the entity types based on the NODE-TYPES specification. In general, we group together entity types that form connected subgraphs of the schema graph shown in Table 1 because they are semantically related. Some of these groups are: (genes, proteins, interactions, pathways, experimentTypes), (drugs, pathways, proteins, interactions, experimentTypes), (genes, interactions, experimentTypes, publications), (diseases, anatomy, phenotype, drugs) and so forth.
   (b) For these groups, we represent the entities as star-shaped structures (the center is the id, and the attributes are the edges) connected by the inter-object relationships in the schema, and create random instances of the entities based on the number of nodes determined in the previous step.
   (c) Once instances of the primary entities are created, we create power-law and community-structured edges following the criteria discussed above.
   (d) We randomly allocate labels from the EDGE-LABELS directive provided no constraints are violated.

3. For nodes allocated to motif composition graphs
   (a) We allocate entity types based on NODE-TYPES as before, but we only consider genes and proteins because they participate in pathways.
   (b) We randomly choose motifs from the motif library as described before and assign labels to these edges so that the labels respect the type required by nodes on both sides of the relationships. Further, some motifs come with their own constraints which are recorded in the motif libraries – for example, motif 7 in Figure 1 needs $C1, C2$ to be proteins whose functions are as enzymes.
4. Now we are left with creating hierarchical graphs on top of the same entity nodes we have used.
   (a) We use a set of rules to determine the type of hierarchy we need for these entity nodes. For example, drugs are connected to drugClass nodes using the `subclassOf` relationship, while anatomy nodes are related to anatomyClass nodes through the `subClassOf` relationship but with several other anatomy nodes using the `partOf` relationship.
   (b) If publications are included in the schema, we create a `cites` hierarchy among them.

6   Sample Test Queries for GDB-H Graphs

In this section we present a sampling of potential test queries against any graph generated from the specification above. The purpose of these queries is to access and retrieve portions of the graph and compute functions on it. We note that there is an important difference between our setting and the TPC-style setting, and most graph benchmarks we have discussed in Section 2. In a TPC-style benchmark, the schema of the benchmark database is completely known and therefore benchmark SQL queries can be posed without difficulty. In a typical graph setting, the benchmarks are essentially algorithmic and hence the structure of the graph is the only object of concern. Through not discussed in Section 2, the RDF/OWL benchmarks like LUBM follow the TPC metaphor – they have a simple schema graph with some inference rules, and their goal is to show the scalability in the size of the data and the ontological inference. In our situation, we have a large heterogeneous graph that use a set of declared entity types, but have a large number of possible connections between these entities – the connnections are based on user-defined guidelines but the connection structure is primarily generated algorithmically. This is notionally similar to the original idea of semistructured data [15] and we will use ideas from semi-structured query languages to develop our test queries. In these queries we use a `select .. from .. where` structure to return nodes, paths and subgraphs. As part of the `select` and `where` clauses we use standard graph operations that any graph database should support.

Q1. *List all genes that are related to some publications, and return the corresponding publications.* A gene can be related to a publication along arbitrary-length paths, but the query does not need to return the paths. We write this query as:

```
select X, Y where genes(X), publications(Y), reachable(Y,X)
```

where `reachable(Y,X)` is a standard graph operation, which reads as Y is reachable from X. We adopt the convention that X will be intantiated by the instances of class `genes`. By our convention not having a `from` clause implies that the query is over the entire graph. However, due to the construction process of the heterogeneous graphs, query will never use hierarchical graph components.

Q2. *Find paths of length upto k that start from proteins, go through phenotypes but not through pathways and end in publications.* This query is parameterized on k. This query is subject to the presence of cycles in motif composition components, and dense parts of the graph. The first problem should be resolved by cycle detection and handling in the path evaluation process, while setting realistic limits on k eliminates the second. We express this query as:

```
select path(p) where proteins(X), pathways(W), phenotypes(H), publications(Y),

        p.first = X, p.last = Y, contains(p,H), not(contains(p,W))
```

where *p* will be a chain of edges. A variant of this query will ask for `select disjoint path(p)`.

Q3. *Find disease pathways having at least 5 connected interaction edges belonging to the same interaction class.* This query shows the equivalent of a "HAVING"' clause in a relational query.

```
select W where pathways(W), Interactions(I), InteractionClass(C),

        subClassOf+(I,C), connected(I), count(I)>=5
```

where `subClassOf+` refers to a chain of `subClassOf` edges from C to I; `connected(I)` returns true if the elements of the set I are connected in the graph.

Q4. *Find the k-neighborhood of the protein with the highest centrality value over all non-hierachical graphs.* This graph query that combines an aggregate function (computing a centrality measure) with a neighborhood query with a user-selected parameter k. We express the query as:

```
select k-neighborhood(X) from components G where protein(X),

    not(type(G), hierarchical), betweenness-centrality(C,X,G); max(C)
```

where `max(C)`, where C is the betweenness-centrality value is logically computed at the end. In this query we use the `from` clause to select a subset of the component graphs for the query, which in this case, is the full graph except the hierarchical components.

Q5. *Find the longest path containing a single edge label.* This aggregate query has a condition on the edge label, and can be expressed as:

```
select path(p) where edge(E), label(L, E), member(E, p),

    not(edge(E'), label(L',E'), member(E',p), L != L'); maxlength(p)
```

where `edge`, `label` and `member` are built-in graph functions, and the function `maxlength` is computed after accumulating the candidate paths. A variant of this query can specify a set of labels and ask for the longest path using only those labels.

Q6. *Find the occurrences of a graph pattern.* The pattern can be purely structural or it can place semantic constraints. The following subgraph extraction query shows a five-edge pattern:

```
select subgraph(G) where edge(E1), edge(E2), edge(E3), edge(E4), edge(E5),
    member([E1,E2,E3,E4,E5],G), E1=(n1 :L1 n2), E2=(n2 :L2 n3), E3=(n1 :L3 n4),
        E4=(n4 :L3 n5), E5=(n2 :L4 n5), L4='is-target-of'
```

where :L1 means a relationship with label L1. Named edges E1 ... E5 are assumed distinct. A shared node variable (n2) represents a shared node. The last predicate represents a semantic constraint on the otherwise structural pattern. These queries do not represent the complete horizon of benchmark queries that can be asked for heterogeneous graphs, but they do represent important classes of questions that go beyond the current one-track approach for graph benchmarks. **Using GDB-H Queries for Benchmarking.** The basic performance metric for the above queries is minimum mean response time and 90th percentile response time as used in the TPC-C benchmark. However, since the data distribution for GDB-H is more complex, we need to be cautious that two runs of the data generator produces comparable data. We envision that the typical use case will be for the user to specify only the mandatory parameters at first; when the system generates data, it will also output the parameters of the remaining distributions used for generating the data. As the user finds the data acceptable, they now use all parameters of the prior run as the guideline to specify the next round of data generation. This will keep the generated data sets comparable.

7   Conclusion and Outlook

In this paper, our intent was to develop three important viewpoints in the context of big data benchmarking. First, graph benchmarks that we see today are mostly geared toward single operations, but they lack "variety", an important component of big data. We proposed a mathematically well-founded heterogenous graph data model and presented its generation methodology with the idea of increasing the variety part of the graph. Second, it is important to go beyond single operations for graph benchmarking and develop query operations that showcase how different graph operators can be effectively combined to formulate and evaluate "larger" queries. We expect our test queries will serve as an initial attempt to create such ad hoc complex queries that make use of different graph functions. Third, our benchmark should really be taken as a "generic template" and while our example application is around biological data, our template can be used for any domain. If the entity types were non-biological objects like person, place, event, geographic object and music, while the relationships connecting them together are like: person × object, visited: person × location, attended: person × event, located-in: event × places and so forth, the system will resemble a social media framework, for example the recently announced Facebook Graph.

Since this is our first attempt to develop a generation algorithm, it can be improved significantly. Currently, edge label assignment is separate from edge assignment – we need to bring them under a single generative model. We must explore some properties of heterogenous graphs to understand its behavior more completely. We must make the generation process faster, possibly using a distributed framework. Finally, the query language must be matured significantly.

References

1. Chakrabarti, D., Zhan, Y., Faloutsos, C.: R-mat: A recursive model for graph mining. In: Proc. 4th SIAM Int. Conf. on Data Mining (2004)
2. Seshadhri, C., Pinar, A., Kolda, T.G.: An in-depth study of stochastic kronecker graphs. In: Proc. of the 11th IEEE Int. Conf. on Data Mining (ICDM), pp. 587–596 (2011)
3. Newman, M.E., Girvan, M.: Mixing patterns and community structure in networks. Statistical Mechanics of Complex Networks, 66–87 (2003)
4. Lancichinetti, A., Fortunato, S., Radicchi, F.: Benchmark graphs for testing community detection algorithms. Phys. Rev. E 78, 04110 (2008)
5. Pham, M.-D., Boncz, P., Erling, O.: S3g2: A scalable structure-correlated social graph generator. In: Nambiar, R., Poess, M. (eds.) TPCTC 2012. LNCS, vol. 7755, pp. 156–172. Springer, Heidelberg (2013)
6. Guo, Y., Pan, Z., Heflin, J.: LUBM: A benchmark for OWL knowledge base systems. J. Web Sem. 3(2-3), 158–182 (2005)
7. Bizer, C., Schultz, A.: The Berlin SPARQL benchmark. Int. J. on Semantic Web and Information Systems (IJSWIS) 5(2), 1–24 (2009)
8. Aiello, W., Chung, F., Lu, L.: A random graph model for power law graphs. Experimental Mathematics 10(1), 53–66 (2001)
9. Seshadhri, C., Kolda, T.G., Pinar, A.: Community structure and scale-free collections of Erdös-Rényi graphs. CoRR abs/1112.3644 (2011)
10. Karrer, B., Newman, M.: Random graph models for directed acyclic networks. Physical Review E 80(4), 046110 (2009)
11. Lima-Mendez, G., van Helden, J.: The powerful law of the power law and other myths in network biology. Mol. BioSyst. 5, 1482–1493 (2009)
12. Chung, F.R.K., Lu, L., Dewey, T.G., Galas, D.J.: Duplication models for biological networks. Journal of Computational Biology 10(5), 677–687 (2003)
13. Kolda, T.G., Pinar, A., Plantenga, T., Seshadhri, C.: A scalable generative graph model with community structure (February 2013),
http://arxiv.org/abs/1302.6636
14. Krumsiek, J., Suhre, K., Illig, T., Adamski, J., Theis, F.J.: Gaussian graphical modeling reconstructs pathway reactions from high-throughput metabolomics data. BMC Systems Biology 5(1), 21 (2011)
15. Quass, D., Rajaraman, A., Sagiv, Y., Ullman, J., Widom, J.: Querying semistructured heterogeneous information. In: Ling, T.W., Mendelzon, A.O., Vieille, L. (eds.) DOOD 1995. LNCS, vol. 1013, pp. 319–344. Springer, Heidelberg (1995)

A Micro-benchmark Suite for Evaluating
HDFS Operations on Modern Clusters*

Nusrat Sharmin Islam, Xiaoyi Lu, Md. Wasi-ur-Rahman,
Jithin Jose, and Dhabaleswar K. (DK) Panda

Department of Computer Science and Engineering, The Ohio State University
{islamn,luxi,rahmanmd,jose,panda}@cse.ohio-state.edu

Abstract. Hadoop Distributed File System (HDFS) is the primary storage system of Hadoop. Many applications use HDFS as the underlying file system due to its portability and fault-tolerance. The most popular benchmark to measure the I/O performance of HDFS is TestDFSIO which involves the MapReduce framework. However, there is a lack of standardized benchmark suite that can help users evaluate the performance of standalone HDFS and make comparisons for different networks and cluster configurations. In this paper, we design and develop a micro-benchmark suite that can be used to evaluate performance of HDFS operations. This paper also illustrates how this benchmark suite can be used to evaluate the performance results of HDFS installations over different networks/protocols and parameter configurations on modern clusters.

Keywords: Big Data, Hadoop, HDFS, Micro-benchmarks, Clusters and Networks.

1 Introduction

In the past several years there has been an immense surge of interest for Big data. Big Data provides ground breaking opportunities for enterprise information management and decision making. As a matter of fact, Big Data fundamentally changes the way decisions are being made in a wide range of domains including health care, biomedical research, internet services, business informatics, scientific computing and others. MapReduce [13] has been proved as a viable model for processing petabytes of data. Hadoop [7] is an open-source implementation of the MapReduce model, and it has gained lots of attentions from academic and industrial communities. The Hadoop software stack contains several middleware components such as Hadoop Distributed File System (HDFS) [30] (filesystem), MapReduce (computation), and HBase [8] (database). Hadoop is derived from Google's MapReduce [13] and Google File System (GFS) [15] which is also the underlying file system of Google's Big Table [11]. As data sizes are steadily increasing, there is an increasing demand for Hadoop to deliver high-performance and scalability continuously. Recent research works [32,27,16,33,20] analyze on the huge performance improvements possible for different cloud computing

---

* This research is supported in part by National Science Foundation grants #OCI-0926691, #OCI-1148371 and #CCF-1213084.

T. Rabl et al. (Eds.): WBDB 2012, LNCS 8163, pp. 129–147, 2014.
© Springer-Verlag Berlin Heidelberg 2014

middlewares using InfiniBand [2] networks. The first ever Hadoop package designed and developed using RDMA over InfiniBand is available for public use from [1]. It provides native InfiniBand Verbs level support for multiple Hadoop components (HDFS, MapReduce and RPC) for Big Data processing and can offer significant performance improvement over the socket-based implementation of Apache Hadoop over InfiniBand.

Hadoop Distributed File System (HDFS) is the primary storage for Hadoop clusters. Both Hadoop MapReduce and HBase rely on HDFS as the underlying basis for providing data distribution and fault tolerance. As the underlying file system, the performance of HDFS operations dramatically influences the performance of the upper layer middlewares, components, and applications. The performance of HDFS operations is determined by many factors related to storage and network configurations in modern clusters, controllable parameters in software (e.g. block-size), data access patterns of applications, and so on. So we often need to tune these factors for the optimal performance based on cluster and workload characteristics. One of the most common tuning methods is to adopt a benchmark tool suite to evaluate the performance metrics in different kinds of system configurations. Currently, the most popular benchmark to measure the I/O performance of HDFS is TestDFSIO [34], which needs to launch the MapReduce framework. According to [21], due to the scheduling delays of the MapReduce framework, HDFS cannot be utilized to its full potential. However, there is a lack of standardized benchmark suite that can help users evaluate the performance of standalone HDFS and make comparisons for different storage, network, and parameter configurations on modern clusters. This kind of benchmarks can also prove to be more relevant for applications using native HDFS (such as HBase) instead of going through the MapReduce layer.

In this paper, we design, develop, and implement a comprehensive micro-benchmark suite to evaluate the performance of standalone HDFS, particularly, the *Read* and *Write* operations. We provide options for varying different benchmark-level parameters such as file size, numbers of concurrent readers for read-only workload, writers for write-only workload and readers and writers for mixed workload. Our benchmark suite can also dynamically set the HDFS configuration parameters like block-size, replication factor, etc. Our benchmarks can also display the HDFS configuration parameters for a Hadoop cluster, as part of the benchmark output and present different statistics like minimum, maximum and average, for the results.

This paper makes the following key contributions:

1. Design, develop, and implement a micro-benchmark suite to evaluate I/O performance of standalone HDFS.
2. Provide a set of standard benchmarks to measure the latency and throughput of HDFS read, write, and mix workload (read and write).
3. Illustrate the performance results of HDFS read and write using our benchmark suite over different networks/protocols and parameter configurations on modern clusters.

The rest of the paper is organized as follows. Section 2 discusses the background of this research. In Section 3, we distinguish our work from existing work in the field. We present our design considerations for the benchmark suite in Section 4 and benchmarks for HDFS operations in Section 5. In Section 6, we show the performance evaluation results. Finally, we conclude in Section 7.

2 Background

2.1 Hadoop Distributed File System (HDFS)

The Hadoop Distributed File System (HDFS) is used as the primary storage for a Hadoop cluster. Figure 1 illustrates the basic architecture of HDFS. An HDFS cluster consists of two types of nodes: NameNode and DataNodes. The NameNode manages the file system namespace. It maintains the file system tree and stores all the meta data. The DataNodes on the other hand, act as the storage system for the HDFS files. HDFS divides large files into blocks of size 64 MB. Each block is stored as an independent file in the local file system of the DataNodes. HDFS usually replicates each block to three (replication factor three) DataNodes. In this way, HDFS guarantees data availability and fault-tolerance.

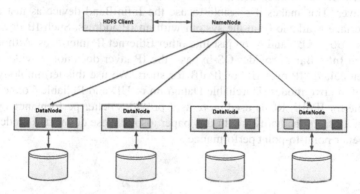

Fig. 1. Overview of HDFS

The HDFS client contacts the NameNode during any kind of file system operations. When the client wants to write a file to HDFS, it gets the block IDs and list of DataNodes for each block from the NameNode. Each block is split into smaller packets and sent to the first DataNode in the pipeline. The first DataNode then replicates each of the packets to the subsequent DataNodes. Packet transmission in HDFS is pipelined; a DataNode can receive from the previous DataNode while it is replicating data to the next DataNode. If the client is running inside a DataNode, then the block is first written to the local file system of the current node. Figure 2(b) illustrates this operation.

For HDFS read, as shown in Figure 2(a), the client first contacts the NameNode to check its access permission and gets the block IDs and locations for each of the blocks. For each block belonging to the file, the client connects with the nearest DataNode and reads the block. Blocks belonging to a particular file are read sequentially by the client.

2.2 High Performance Networks

In this section, we present an overview of the different networking technologies that can be utilized in data center for high-performance communication. During the past decade, the field of High-Performance Computing (HPC) has been witnessing a transition to commodity clusters with modern interconnects such as InfiniBand and 10Gigabit Ethernet.

InfiniBand. InfiniBand [18] is an industry standard switched fabric that is designed for interconnecting nodes in HPC clusters. It is a high-speed, general purpose I/O interconnect that is widely used by scientific computing centers world-wide. The recently released TOP500 rankings in June 2013 indicate that InfiniBand technology now provides interconnects on 226 systems (45.2%) and it is the most-used internal system interconnect technology. One of the main features of InfiniBand is Remote Direct Memory Access (RDMA). This feature allows software to remotely read or update memory contents of another remote process without any software involvement at the remote side. InfiniBand has started making inroads into the commercial domain with the recent convergence around RDMA over Converged Enhanced Ethernet (RoCE) [31].

InfiniBand software stacks, such as OpenFabrics [23], provide driver for implementing the IP layer. This makes it possible to use the InfiniBand device as just another network interface available from the system with an IP address. Such IB devices are presented as ib0, ib1 and so on just like other Ethernet IP interfaces. Although the verbs layer in InfiniBand provides OS-bypass, the IP layer does not provide so. This layer is often called "IP-over-IB" or IPoIB for short. We use this terminology in the paper. Out of the two modes (Unreliable Datagram or UD and Reliable Connection or RC) available for IPoIB, RC is used more as it provides better performance by leveraging reliability from the hardware. In this paper also, we use connected mode IPoIB, which has better point-to-point performance.

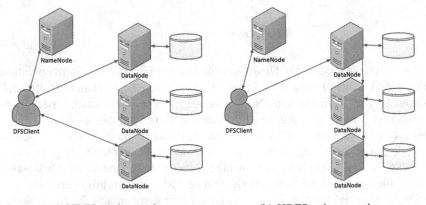

(a) HDFS read operation   (b) HDFS write operation

Fig. 2. HDFS read and write operation

10 Gigabit Ethernet. In data center environments for achieving better performance with respect to higher bandwidth, 10 Gigabit Ethernet is typically used. It is also realized that traditional sockets interface may not be able to support high communication rates [9]. Towards that effort, iWARP (Internet Wide Area RDMA Protocol) standard was introduced for performing RDMA over TCP/IP [28]. The iWARP semantics is very similar to the verbs layer used by InfiniBand, with the exception of requiring a connection manager. In fact, the OpenFabrics [23] network stack provides a unified interface for both iWARP and InfiniBand.

In this study, we have used 1 GigE, 10 GigE and IPoIB interconnects in the experiments.

2.3 Solid State Drive (SSD) Overview

Solid State Drives have amassed a lot of attention over the recent past owing to significant data-throughput and efficiency gains over traditional spinning-disks. Although it is a mere physical array of fast flash-memory packages, the core-intelligence of an SSD can be attributed to its Flash Translation Layer (FTL) which plays a vital role in the adoption of this technology. Some of the major functionalities of an SSD, such as Wear Leveling, Garbage Collection and Logical Block Mapping are packed into the FTL. High bandwidth as well as low latency makes SSD an ideal candidate to be used in the DataNodes of the Hadoop cluster in order to lessen the I/O bottlenecks for HDFS applications.

3 Related Work

Benchmarking is important for evaluating Big Data systems, and extensive work has been done in this area. Hadoop [7] contains a set of built-in micro-benchmarks such as TeraSort [24], Sort [4], Word Count [6], RandomWriter [3], TestDFSIO [34], etc. TestDFSIO, the most popular HDFS benchmark, is implemented as a Hadoop MapReduce job. Each map task in TestDFSIO opens an HDFS file to write or read sequentially and measures the data size and execution time. A single reduce task aggregates the performance results of all the map tasks by computing the average I/O rate and throughput. J. Shafer et al. [21] analyzed HDFS performance using the TestDFSIO program and revealed that the bottleneck could be the MapReduce tasks scheduling in Hadoop, the HDFS Java implementation, and the native disk I/O scheduling.

MRBench [22] executes highly complex queries on large amount of relational data and provides micro-benchmarks in the form of MapReduce jobs of TPC-H [5]. MRBS [29] is a benchmark suite to evaluate the dependability of MapReduce systems, and it includes five benchmarks for several application domains and a wide range of execution scenarios. The authors of HiBench [17] have extended the DFSIO program to compute the aggregated bandwidth by disabling the speculative execution of the MapReduce framework. HiBench also evaluates Hadoop in terms of system resource utilization (e.g. CPU and memory). MalStone [10] is a benchmark suite designed to measure the performance of cloud computing middleware when building data mining models. Yahoo! Cloud Serving Benchmark (YCSB) [12] is a set of benchmarks for performance evaluations of key/value-pair and cloud data serving systems. YCSB++ [25]

further extends YCSB to improve performance understanding and debugging. The authors in [14] have compared SQL Server and MongoDB on interactive data-serving environments using the YCSB benchmark. The authors in [26] have performed comprehensive evaluations of six open-source data stores as part of application performance monitoring for Big Data. Traditional benchmark suite for POSIX file system, IOZone [19], generates and tests a variety of file system operations.

By analysis on these related works, we find that there is a lack of a suite of micro-benchmarks to evaluate the native HDFS operations and current research works were not able to analyze the variable HDFS performance on different hardware. Our proposed benchmark suite addresses such shortcomings in the Big Data community for evaluating HDFS performance. Our benchmark suite does not need to launch any other job (such as MapReduce) and can be used in a simple manner to evaluate the I/O performance of standalone HDFS. As a result, our benchmark suite can be used to carry out performance comparison of HDFS for different storage, network, protocol, and parameter configurations on modern clusters. This suite of micro-benchmarks is particularly designed for HDFS and can also be applied for other distributed file systems that have similar APIs as HDFS.

4   Design Considerations for the Benchmark Suite

Among the operations of HDFS, the most important ones are sequential write, sequential read and random read. HDFS performance is usually measured by the latency and throughput of these operations. The performance of HDFS is influenced by a range of factors such as underlying network as well as storage, HDFS configuration parameters and data access patterns. We consider the aspects described below designing the benchmark suite.

Network: The performance of HDFS operations is influenced by the underlying interconnect or protocol to a great extent [20,32]. During HDFS *Write*, data packets are replicated from one DataNode to another along the pipeline. Therefore, high performance networks can speed up the replication process for data-intensive applications. Faster interconnects can enhance the performance of HDFS *Read* also. If the HDFS client co-exists with a block replica in the same DataNode, the client simply reads the block locally. But, if the DFSClient runs in a separate node outside the Hadoop cluster, read performance can be improved by the usage of faster networks and protocols. Besides, the optimal packet-size for HDFS varies with the interconnect or protocol [20]. HDFS performance also depends on HDFS block-size. The DFSClient communicates with the NameNode before it reads or writes an HDFS block. For any particular file size, larger block-size results in reduced number of communications with the NameNode. Therefore, HDFS block-size is an important parameter in determining the optimal performance for HDFS applications. Our benchmark suite facilitates easy configuration of these parameters for different network types, to help understand the interaction between network characteristics and HDFS performance.

Storage: The number and type (HDD, SSD or combination) of the underlying storage device can have significant impact on HDFS performance. The optimal values of the configuration parameters may also vary with the storage type. Therefore, our benchmark suite, equipped with options to change HDFS configuration parameters from user-level input, makes the performance characterization over different storage platforms easier and user friendly.

HDFS Configuration Parameters: The performance of HDFS largely depends on various configuration parameters like replication factor, HDFS file I/O buffer size, etc. In our benchmarks, we provide options to set these HDFS configuration parameters dynamically with the values provided by the users. If no values are provided from the user level, the benchmark will run with the parameters specified in the HDFS configuration file.

Data Access Patterns: The performance of HDFS operations is also influenced by different data access patterns of workloads. In our benchmarks, we focus on three kinds of data access patterns: Sequential, Random, and Mixed (user specified ratio). Users can select the workload type by a simple parameter.

5 Benchmarks for HDFS Operations

In this study, we develop a micro-benchmark suite for HDFS. Each of the benchmarks is written in Java. We design and implement the following set of benchmarks:

Sequential Write Latency (HDFS-SWL): This benchmark takes the file name and size as inputs and outputs the total time to write this file to HDFS. HDFS write is performed sequentially by dividing the file into a set of blocks. For this, the benchmark invokes the HDFS `create()` API to get an instance of `FSDataOutputStream`. Data bytes are then written to HDFS by using the `write()` method of `FSDataOutputStream`. The benchmark starts a timer just after creating the file and stops this after the file is closed. The time measured in this way is reported as the latency of sequential write for the file specified by the user. The pseudocode of the benchmark is presented in Algorithm 1.

Sequential or Random Read Latency (HDFS-SRL or HDFS-RRL): This benchmark takes the file name, size, access pattern (random or sequential) and seek interval (for random only) as inputs and outputs the time to read the file from HDFS. For this, the benchmark invokes the HDFS `open()` API to get an instance of `FSDataInputStream`. Data bytes are then read from HDFS by using the sequential or random `read()` method of `FSDataInputStream`. The benchmark starts a timer just after opening the file and stops this after read completion. The time measured in this way is reported as the read latency for the data size specified by the user.

Sequential Write Throughput (HDFS-SWT): The user can input the number of concurrent writers and the size of data in (MB) per writer. The benchmark outputs the throughput per writer by dividing the data size with the write-time required by it. The total throughput is calculated by multiplying the average throughput per writer with the number of writers.

In order to launch multiple HDFS clients (writers) at the same time, we have designed a job-launcher. The job-launcher is a Java program that starts the writer processes in different nodes. It also aggregates the throughput values from different writers and outputs the total write throughput.

Input: File Name $fName$, File Size $fSize$, Integer $iterationCount$, A
       sequence of configuration parameters, $C = c_1, c_2, \ldots, c_n$
Result: Latency for Sequential Write
create a Hadoop Configuration object
if $C.length! = 0$ **then**
  | set conf parameters
else
  | load default parameters
end
while $iterationCount > 0$ **do**
  | create FSDataOutputStream object $fsOut$
  | start timer
  | call $fsOut.write(fName, fSize)$
  | $fsOut.close()$
  | end timer
  | delete $fsOut$ from HDFS
end
Print Average Latency

Algorithm 1. PSEUDOCODE OF SEQUENTIAL WRITE LATENCY (HDFS-SWL) BENCHMARK

Sequential Read Throughput (HDFS-SRT): This benchmark works in a similar manner as the one for write workload. Here the inputs are number of concurrent readers and read size per reader. It calculates the total throughput for sequential read. In this case also, the Java-based job-launcher aggregates the read throughput from different readers and outputs the total throughput.

Sequential Read-Write Throughput (HDFS-SRWT): This benchmark calculates the total throughput when HDFS read and write are occurring simultaneously. The benchmark takes the numbers of readers, writers and size per reader and writer as inputs. The read/write ratio can be varied by varying the number of concurrent readers and writers and also the data size for each.

In all these benchmarks, the users can also provide different HDFS configuration parameters as input as discussed in Section 4. Table 1 lists the parameters of our benchmark suite. Each benchmark can report the configuration parameters in use for it as part

Table 1. Benchmark parameter list

Benchmark	File Name	File Size	HDFS Parameters	Readers	Writers	Random/Seq Read	Seek Interval
HDFS-SWL	✓	✓	✓				
HDFS-SRL/RRL	✓	✓	✓			✓	✓(RRL)
HDFS-SWT		✓	✓		✓		
HDFS-SRT		✓	✓	✓			
HDFS-SRWT		✓	✓	✓	✓		

of the output. The benchmark suite also calculates statistics like Min, Max, and Avg. latency and throughput.

6   Performance Evaluation

In this section, we present the detailed performance evaluations of HDFS using our micro-benchmark suite.

6.1   Experimental Setup

We have used two different cluster configurations.

(1) **Intel Westmere Cluster (Cluster A)**: This cluster consists of 160 compute nodes with Intel Westmere series of processors using Xeon Dual quad-core processor nodes operating at 2.67 GHz with 12GB RAM and 160GB HDD. Each node is equipped with MT26428 QDR ConnectX HCAs (32 Gbps data rate) with PCI-Ex Gen2 interfaces. The nodes are interconnected using a Mellanox QDR switch. Each node runs Red Hat Enterprise Linux Server release 6.1 (Santiago) at kernel version 2.6.32-131 with OpenFabrics version 1.5.3.

(2) **Intel Westmere Cluster with Larger Memory (Cluster B)**: Nodes in this cluster have the same configurations as Cluster A but with 24GB of RAM each. Additionally, twelve of the storage nodes are equipped with three 1TB HDD each and the rest four nodes have 300 GB OCZ VeloDrive PCIe SSD. Four of the storage nodes also have NetEffect NE020 10Gb Accelerated Ethernet Adapter (iWARP RNIC) that are connected using a 24 port Fulcrum Focalpoint switch.

6.2   Evaluations over different Interconnects and Protocols:

In this section, we discuss the performance of our benchmarks over different interconnects and protocols like 1 GigE, 10 GigE, and IPoIB. HDFS replication factor is *three*, block-size is 64 MB, file io-buffer-size is 4 KB, and packet-size is 64 KB for GigE networks and 128 KB for IPoIB. These packet-sizes are found to be optimal for the corresponding interconnects [20].

HDFS-SWL, HDFS-SRL, HDFS-RRL: In these experiments, the HDFS NameNode and client run exclusively on two different nodes. Since HDFS write is more network intensive (as it involves replication), we have performed experiments with HDFS write in four and 32 DataNodes in Cluster A in order to observe how the write performance vary with number of DataNodes. Each of the DataNodes has single disk per node.

Figures 3(a) and 3(b) show the latency of file write using SWL in four and 32 DataNodes, respectively. For the same file size, the latency decreases as we move from four to 32 DataNodes. And the performance improvement of IPoIB is bigger than that of 1 GigE. This is because the same amount of data is distributed to more number of disks when 32 DataNodes are used. High performance network (like InfiniBand) and protocol (like IPoIB) can achieve more speed up when the I/O bottleneck is reduced or eliminated by using more disks.

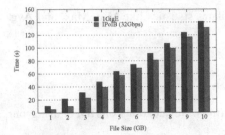
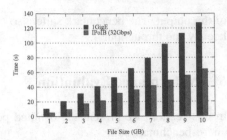

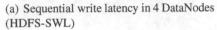

(a) Sequential write latency in 4 DataNodes (HDFS-SWL)

(b) Sequential write latency in 32 DataNodes (HDFS-SWL)

Fig. 3. HDFS write latency in Cluster A

Figures 4(a) and 4(b) illustrate the performance results with HDFS-SRL, HDFS-RRL, respectively, for file sizes 1 GB-10 GB. We have used a seek interval of 500 for HDFS-RRL. For the same read size, the latency of HDFS-RRL is slightly higher than that of HDFS-SRL. The latency of random read increases with seek interval.

Figures 5(a) and 5(b) show the HDFS-SWL and HDFS-SRL latency results over 1 GigE, IPoIB (32Gbps) and 10 GigE. In this experiment, we use four DataNodes in Cluster B. Each of the DataNodes uses single HDD per node as HDFS data directory. The client (reader/writer) runs in a remote node. From the results, it is observed that, 10 GigE provides smaller write latency than that of 1 GigE while IPoIB (32Gbps) gives the smallest latency in this configuration.

HDFS-SWT: Figures 6(a) and 6(b) show the performance results of SWT for four and eight writers, respectively. The file size varies from 1 GB-10 GB. From the figures we observe that the total throughput increases as the number of writers is increased from four to eight. Also for IPoIB, the throughput decreases as the file size increases. This is because, for high performance network protocol (like IPoIB), the bottleneck moves from network to I/O as the file size increases. Larger file sizes increase the amount of I/O which results in reduced throughput for IPoIB with increasing file sizes.

We have performed similar experiments using a 32-DataNode Hadoop cluster. In this case, the throughput does not decrease significantly with increasing file sizes. Because the same amount of data is now distributed to more number of nodes which in turn reduces the I/O bottleneck by placing more data in disk cache and thus achieves higher throughput. Figures 7(a) and 7(b) show the results.

Figure 8(a) shows the HDFS-SWT write throughput results over 1 GigE, IPoIB (32Gbps) and 10 GigE. In this experiment, the four writers run in four different DataNodes in Cluster B. Each of the DataNodes has single HDD per node. From the results, it is observed that, 10GigE provides better write throughput than that of 1 GigE while IPoIB (32Gbps) gives the highest throughput in this configuration.

Also for 10 GigE and IPoIB, the throughput for smaller file size (i.e. 5 GB) is comparatively much higher than that of the others. This is because, for smaller file sizes, most of the data is written to the disk cache; thus the amount of data actually going to disk is very small. Therefore, the throughputs are quite high.

HDFS-SRT: Figure 8(b) shows the HDFS-SRT read throughput results over 1 GigE, IPoIB (32Gbps) and 10 GigE. In this experiment, the four readers run in a remote node

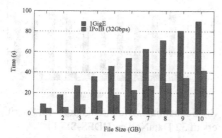
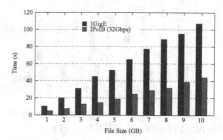

(a) Sequential read latency in 4 DataNodes (HDFS-SRL)

(b) Random read latency in 4 DataNodes (HDFS-RRL)

Fig. 4. HDFS read latency in Cluster A

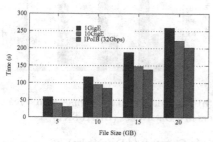
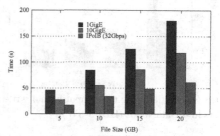

(a) Sequential write latency in 4 DataNodes (HDFS-SWL)

(b) Sequential read latency in 4 DataNodes (HDFS-SRL)

Fig. 5. HDFS write and read latency in Cluster B

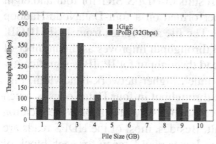
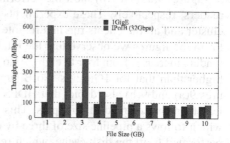

(a) Sequential write throughput with 4 writers in 4 DataNodes (HDFS-SWT)

(b) Sequential write throughput with 8 writers in 4 DataNodes (HDFS-SWT)

Fig. 6. HDFS write throughput in Cluster A (4 DataNodes)

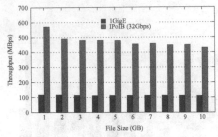

(a) Sequential write throughput with 4 writers in 32 DataNodes (HDFS-SWT)

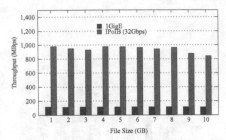

(b) Sequential write throughput with 8 writers in 32 DataNodes (HDFS-SWT)

Fig. 7. HDFS write throughput in Cluster A (32 DataNodes)

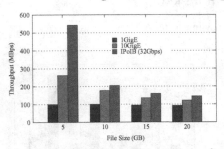

(a) Sequential write throughput with 4 writers in 4 DataNodes (HDFS-SWT)

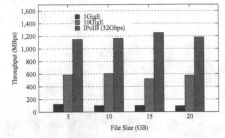

(b) Sequential read throughput with 4 readers in 4 DataNodes (HDFS-SRT)

Fig. 8. HDFS write and read throughput in Cluster B (4 DataNodes)

outside the Hadoop cluster. From the results, it is observed that, 10GigE provides better read throughput than that of 1 GigE while IPoIB (32Gbps) gives the highest throughput in this configuration. From all the results of HDFS-SRT, it is evident that, the read throughput does not vary much with increasing file sizes. Most of the HDFS reads occur from the cache, which results in higher read throughput.

Figures 9(a) and 9(b) show the performance results of HDFS-SRT for four and eight readers, respectively. These experiments are performed on a four-DataNode Hadoop cluster and all the clients run in a remote node. The file size varies from 1 GB-10 GB. From the figures we observe that the total read throughput improves as the number of readers increases from four to eight. Also the read throughput for IPoIB is significantly higher than that for 1GigE as IPoIB supports much higher network bandwidth. Figures 9(a) and 9(b) also depict that HDFS write throughput in four DataNodes is much less than that of read throughputs. This is because HDFS write involves replication which follows a pipeline size of three by default. Besides, most of the HDFS reads are performed from the disk cache, which results in lower read latency for each block of the file, thus providing a higher read throughput.

HDFS-SRWT: The performance results of average latency and total throughput of HDFS-SRWT are illustrated in Figures 10(a) and 10(b), respectively. Here, we have used a workload of 50% read-50% write; each reader and writer works on equal amount of data. The average read/write latency decreases in HDFS-SRWT compared to that in

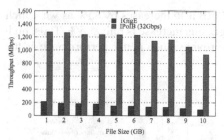

(a) Sequential read throughput with 4 readers in 4 DataNodes (HDFS-SRT)

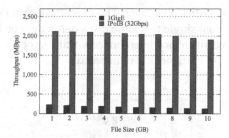

(b) Sequential read throughput with 8 readers in 4 DataNodes (HDFS-SRT)

Fig. 9. HDFS read throughput in Cluster A (4 DataNodes)

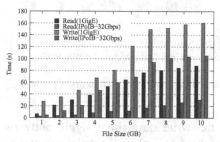

(a) Read and write latency with 4 readers and 4 writers (HDFS-SRWT)

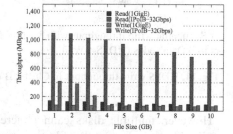

(b) Read and write throughput with 4 readers and 4 writers (HDFS-SRWT)

Fig. 10. HDFS mixed workload evaluation in Cluster A

HDFS-SWL/HDFS-SRL as the same amount of data is being read/written by larger number of reader/writer threads. The throughputs also decrease compared to those in HDFS-SRT/HDFS-SWT due to the contention among reader and writer threads.

6.3   Evaluations over different Storage Platforms:

In this section, we discuss the performance of our benchmarks over different storage platforms.

HDFS-SWT: Our experiments reveal that the performance of HDFS write is influenced by an interesting interplay between network and storage. We run HDFS-SWT benchmark with eight and sixteen clients in an eight-DataNode Hadoop cluster and each of the clients run in a DataNode. For this, we use eight storage nodes in Cluster B as the DataNodes. Each of these nodes is equipped with two HDDs per node. From Figures 11(a) and 11(b), we observe that, for high performance networks like IPoIB, the throughput is significantly increased for larger file sizes as we switch from single to double disks per node. Networks like 1GigE do not show the similar trend. As for 1GigE, the limited network bandwidth itself is the bottleneck. Whereas, the higher bandwidth of IPoIB helps to increase the write throughput further, as the I/O bottleneck is eliminated by installing multiple disks per node. Also for 30GB file size, the gain in throughput for IPoIB compared to 1GigE is 93% over single disk. This gain increases up to 239% for two disks per node. The write throughput of IPoIB also increases by 85% as we switch from single to double disks per node. For 1GigE, this increase is only 5.5%. Thus, high performance interconnects and protocols can provide much better performance as the I/O bottleneck is eliminated.

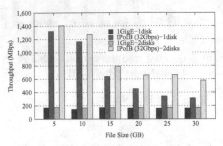

(a) Sequential write throughput with 8 writers in 8 DataNodes (HDFS-SWT)

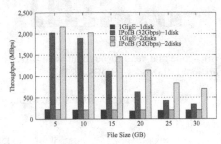

(b) Sequential write throughput with 16 writers in 8 DataNodes (HDFS-SWT)

Fig. 11. HDFS write throughput in Cluster B (8 DataNodes, single vs double disks per node)

However, multiple disks cannot increase the write throughput significantly for smaller file sizes. For smaller file sizes, most of the data is placed in disk cache during file write. Thus, more disks cannot help improve the performance much in this case.

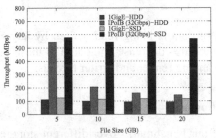

(a) Sequential write throughput with 4 writers in 4 DataNodes (HDFS-SWT)

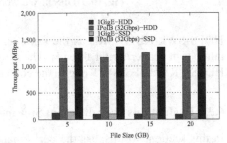

(b) Sequential read throughput with 4 readers in 4 DataNodes (HDFS-SRT)

Fig. 12. HDFS write and read throughput in Cluster B (4 DataNodes, HDD vs SSD per node)

Figures 12(a) and 12(b) show the comparison of throughputs of HDFS-SWT and HDFS-SRT over HDD and SSD. In the HDFS-SWT test, we have used four writers each running in a DataNode. As observed from Figure 12(a), for high performance networks like IPoIB, the write throughput in SSD is much higher than that in HDD for larger file sizes. This is because of the higher write bandwidth of the SSDs. But networks like 1 GigE fail to utilize the improved write bandwidth of SSDs due to its limited network bandwidth. Thus, high performance interconnects and protocols can utilize the benefit of improved storage and offer much better performance. However, for smaller file sizes, the throughput does not depend much on the storage as most of the data is placed inside disk cache during HDFS write.

HDFS-SRT: Figure 12(b) shows the read throughputs of HDFS-SRT on HDD and SSD platforms. As we observe, due to the improved read bandwidth of SSDs, the read throughput increases on this platform. However, the increase is not as significant as in case of HDFS write. The reason behind this is, during HDFS write, lots of threads including the replication threads access the same disk which degrades the throughput. But for read, the number of concurrent threads accessing the disk is much less. The read operation involves much less contention than write. Thus, the increase in read throughput is less compared to that in write throughput with improved storage.

We have also performed HDFS-SRT experiments with eight readers in an eight-DataNode Hadoop cluster. Each client runs in a DataNode. The client first writes a file to HDFS and then reads it back. As observed from Figure 13(a), the read throughput does not depend on the underlying network in this case, as most of the blocks will be read locally from the same node in which the client is running. The read throughputs are also quite high irrespective of the network due to the data prefetching done during HDFS read.

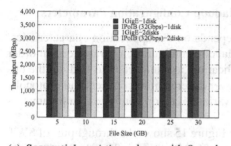 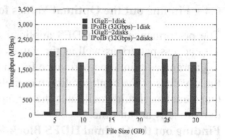

(a) Sequential read throughput with 8 readers in 8 DataNodes (HDFS-SRT)

(b) Sequential read throughput with 8 readers in a remote node (HDFS-SRT)

Fig. 13. HDFS read throughput in Cluster B (8 DataNodes, single vs double disks per node)

Figure 13(b) shows the read throughputs when eight readers run in a remote node outside the Hadoop cluster. In this case, the throughput increases for high performance networks like IPoIB, as the HDFS blocks are read by clients running in a remote node. Thus, the read latency decreases for IPoIB which, in turn, causes the throughput to increase.

From both Figures 13(a) and 13(b), we observe that, HDFS read performance does not vary much as we switch from single to double disks per node. This is because, the HDFS blocks belonging to a file are distributed to different DataNodes in the cluster during replication. While reading, different readers read the blocks from different nodes. As a result, the contention among the readers is less than that in HDFS write (as write involves replication also). Therefore, multiple disks cannot improve the read performance significantly.

HDFS-SRWT: We have performed HDFS-SRWT experiments with eight readers and eight writers in an eight-DataNode Hadoop cluster. The readers and writers run in a

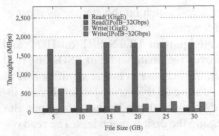
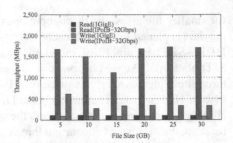

(a) Read and write throughput with 8 readers and 8 writers with single disk per node (HDFS-SRWT)

(b) Read and write throughput with 8 readers and 8 writers (SRWT) with two disks per node (SRWT)

Fig. 14. HDFS throughput for mix workload in Cluster B (8 DataNodes, single vs double disks per node)

remote node and we used a workload of 50% read-50% write. As can be seen from Figures 14(a) and 14(b) multiple disks can increase the write throughput for HDFS.

6.4   Finding out the Optimal Values for Hadoop Configuration Parameters:

The performance of an HDFS application depends on the application characteristics, data access pattern as well as the values of Hadoop configuration parameters. In order to guarantee optimal performance for an application, it is important to find out the optimal values of the configuration parameters over different interconnects and protocols. Our benchmarksuite facilitates the tuning of various Hadoop configuration parameters dynamically at run-time.

Finding out the Optimal HDFS Block-Size: Figure 15 shows the throughputs of SWT for different HDFS block-sizes of 128MB and 256MB. Figure 8(a) shows the write throughput for HDFS block-size of 64MB. As observed from these figures, HDFS write throughput is maximized for the block-size of 128MB in our platform. A file has fewer blocks if the block size is larger. This makes it possible for the client to read/write more data with less interaction with the Namenode. It also reduces the total size of metadata in the Namenode (this can be an important consideration for extremely large file systems). In order to maximize the throughput for a large file, it is better to use a larger HDFS block-size. On the other hand, smaller block-size is more suitable for smaller file sizes.

Effect of HDFS Replication Factor on Throughput: Figure 16 shows the write throughputs of SWT for varying replication factors. Figure 8(a) shows the write throughput for HDFS replication factor of three. As observed from these figures, HDFS write throughput is maximized for a replication factor of one and with this replication factor, the throughput does not depend on the interconnect or protocol since most of the data is written to the node in which the writer is running. The throughputs of SWT are better for smaller replication factor. Thus, SWT provides better throughput for replication factor of two compared to that of three. This is because, HDFS performs a pipelined replication. The higher the length of pipeline, the longer the latency to complete the replication process, which, in turn, reduces the throughput. HDFS replication factor is

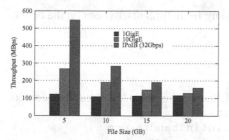

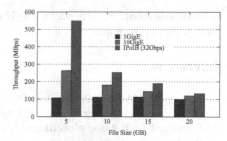

(a) Sequential write throughput with 4 writers in 4 DataNodes with HDFS block-size of 128MB(HDFS-SWT)

(b) Sequential write throughput with 4 writers in 4 DataNodes with HDFS block-size of 256MB (HDFS-SWT)

Fig. 15. HDFS write throughput in Cluster B (4 DataNodes)

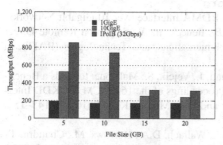

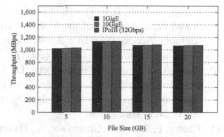

(a) Sequential write throughput with 4 writers in 4 DataNodes with replication factor of 2 (HDFS-SWT)

(b) Sequential write throughput with 4 writers in 4 DataNodes with replication factor of 1 (HDFS-SWT)

Fig. 16. HDFS write throughput in Cluster B (4 DataNodes)

a per file option. Depending on the requirement of the application, users can increase or decrease the replication factor of each specific file in the Hadoop cluster.

7 Conclusion and Future Work

In this paper, we have designed, developed and implemented a micro-benchmark suite to evaluate performance of standalone HDFS. We have provided benchmarks for measuring HDFS *Read* and *Write* latencies. Our benchmarks can also measure the throughputs for read-only, write-only and mixed workloads. We have designed a flexible infrastructure for the benchmarks such that the values of different HDFS configuration parameters can be set dynamically. As an illustration, we have presented performance results of our benchmarks for HDFS over different interconnects on modern cluster.

The benchmark suite can prove to be helpful for designing and evaluating applications that invoke HDFS directly without going through the MapReduce layer. We plan to make this micro-benchmark suite available to the Big Data community via an opensource release in future.

Acknowledgment. We are indebted to Hao Wang for helpful discussions on the benchmark design in the paper.

References

1. Hadoop-RDMA: High-Performance Design of Hadoop over RDMA-enabled Interconnects, `http://hadoop-rdma.cse.ohio-state.edu/`
2. InfiniBand Trade Association, `http://www.infinibandta.com`
3. RandomWriter, `http://wiki.apache.org/hadoop/RandomWriter`
4. Sort, `http://wiki.apache.org/hadoop/Sort`
5. TPC Benchmark H - Standard Specication, `http://www.tpc.org/tpch`
6. WordCount, `http://wiki.apache.org/hadoop/WordCount`
7. Apache Hadoop, `http://hadoop.apache.org/`
8. Apache HBase, `http://hbase.apache.org`
9. Balaji, P., Shah, H.V., Panda, D.K.: Sockets vs RDMA Interface over 10-Gigabit Networks: An In-depth Analysis of the Memory Traffic Bottleneck. In: Workshop on Remote Direct Memory Access (RDMA): Applications, Implementations, and Technologies (RAIT), in Conjunction with IEEE Cluster (2004)
10. Bennett, C., Grossman, R.L., Locke, D., Seidman, J., Vejcik, S.: Malstone: towards a Benchmark for Analytics on Large Data Clouds. In: Proceedings of the 16th ACM SIGKDD International Conference on Knowledge Discovery and Data Mining, KDD 2010, pp. 145–152. ACM, New York (2010)
11. Chang, F., Dean, J., Ghemawat, S., Hsieh, W.C., Wallach, D.A., Burrows, M., Chandra, T., Fikes, A., Gruber, R.: Bigtable: A Distributed Storage System for Structured Data. In: The Proceedings of the Seventh Symposium on Operating System Desgin and Implementation (OSDI 2006), WA (November 2006)
12. Cooper, B.F., Silberstein, A., Tam, E., Ramakrishnan, R., Sears, R.: Benchmarking Cloud Serving Systems with YCSB. In: The Proceedings of the ACM Symposium on Cloud Computing (SoCC 2010), Indianapolis, Indiana, June 10-11 (2010)
13. Dean, J., Ghemawat, S.: MapReduce: Simplified Data Processing on Large Clusters. In: OSDI 2004: Proceedings of the 6th conference on Symposium on Opearting Systems Design and Implementation. USENIX Association (2004)
14. Floratou, A., Teletia, N., DeWitt, D.J., Patel, J.M., Zhang, D.: Can the Elephants Handle the NoSQL Onslaught? In: The Proceedings of the VLDB Endowment, VLDB 2012 (2012)
15. Ghemawat, S., Gobioff, H., Leung, S.: The Google File System. In: The Proceedings of the 19th ACM Symposium on Operating Systems Principles (SOSP 2003), NY, USA, October 19-22 (2003)
16. Huang, J., Ouyang, X., Jose, J.: High-Performance Design of HBase with RDMA over InfiniBand. In: IEEE Int'l Parallel and Distributed Processing Symposium, IPDPS 2011 (May 2011)
17. Huang, S., Huang, J., Dai, J., Xie, T., Huang, B.: The HiBench Benchmark Suite: Characterization of the MapReduce-based Data Analysis. In: IEEE 26th International Conference on Data Engineering Workshops, ICDEW (2010)
18. Infiniband Trade Association, `http://www.infinibandta.org`
19. IOzone, `http://www.iozone.org/`
20. Islam, N.S., Rahman, M.W., Jose, J., Rajachandrasekar, R., Wang, H., Subramoni, H., Murthy, C., Panda, D.K.: High Performance RDMA-based Design of HDFS over InfiniBand. In: The International Conference for High Performance Computing, Networking, Storage and Analysis, SC (November 2012)

21. Shafer, J., Cox, S.R.: A.L.: The Hadoop Distributed Filesystem: Balancing Portability and Performance. In: The Proceedings of the IEEE International Symposium on Performance Analysis of Systems and Software (ISPASS 2010), White Plains, NY, March 28-30 (2010)

22. Kim, K., Jeon, K., Han, H.: MRBench: A Benchmark for MapReduce Framework. In: 14th IEEE International Conference on Parallel and Distributed Systems, ICPADS 2008, pp. 11–18 (2008)

23. OpenFabrics Alliance, http://www.openfabrics.org/

24. Owen, O'Malley: Terabyte sort on apache hadoop, http://sortbenchmark.org/Yahoo-Hadoop.pdf

25. Patil, S., Polte, M., Ren, K., Tantisiriroj, W., Xiao, L., López, J., Gibson, G., Fuchs, A., Rinaldi, B.: YCSB++: Benchmarking and Performance Debugging Advanced Features in Scalable Table Stores. In: Proceedings of the 2nd ACM Symposium on Cloud Computing, SOCC 2011, pp. 9:1–9:14. ACM, New York (2011)

26. Rabl, T., Sadoghi, M., Jacobsen, H.-A., Gómez-Villamor, S., Muntés-Mulero, V., Mankovskii, S.: Solving Big Data Challenges for Enterprise Application Performance Management. In: The Proceedings of the VLDB Endowment, VLDB 2012 (2012)

27. Rahman, M.W., Huang, J., Jose, J., Ouyang, X., Wang, H., Islam, N., Subramoni, H., Murthy, C., Panda, D.K.: Understanding the Communication Characteristics in HBase: What are the Fundamental Bottlenecks? In: IEEE International Symposium on Performance Analysis of Systems and Software, ISPASS (April 2012)

28. RDMA Consortium: Architectural Specifications for RDMA over TCP/IP, http://www.rdmaconsortium.org/

29. Sangroya, A., Serrano, D., Bouchenak, S.: MRBS: Towards Dependability Benchmarking for Hadoop Mapreduce. In: Caragiannis, I., et al. (eds.) Euro-Par Workshops 2012. LNCS, vol. 7640, pp. 3–12. Springer, Heidelberg (2013)

30. Shvachko, K., Kuang, H., Radia, S., Chansler, R.: The Hadoop Distributed File System. In: IEEE 26th Symposium on Mass Storage Systems and Technologies, MSST (2010)

31. Subramoni, H., Lai, P., Luo, M., Panda, D.K.: RDMA over Ethernet - A Preliminary Study. In: Proceedings of the 2009 Workshop on High Performance Interconnects for Distributed Computing, HPIDC 2009 (2009)

32. Sur, S., Wang, H., Huang, J., Ouyang, X., Panda, D.K.: Can High Performance Interconnects Benefit Hadoop Distributed File System? In: Workshop on Micro Architectural Support for Virtualization, Data Center Computing, and Clouds, in Conjunction with MICRO 2010, Atlanta, GA (2010)

33. Wang, Y., Que, X., Yu, W., Goldenberg, D., Sehgal, D.: Hadoop Acceleration through Network Levitated Merge. In: Proceedings of 2011 International Conference for High Performance Computing, Networking, Storage and Analysis, SC 2011 (2011)

34. White, T.: Hadoop: The Definitive Guide. O'Reilly Media, Inc. (October 2010)

Assessing the Performance Impact
of High-Speed Interconnects on MapReduce*

Yandong Wang, Yizheng Jiao, Cong Xu, Xiaobing Li, Teng Wang, Xinyu Que,
Cristi Cira, Bin Wang, Zhuo Liu, Bliss Bailey, and Weikuan Yu

Department of Computer Science
Auburn University
{wangyd,yzj0018,congxu,xbli,tzw0019,xque,
cmc0031,bwang,zhuoliu,wkyu,bailebn}@auburn.edu

Abstract. Hadoop is a successful open-source implementation of MapReduce
programming model. It has been widely adopted by many leading industry companies for big data analytics. However, its intermediate data shuffling is a time-
consuming operation that impacts the total execution time of MapReduce
programs. Recently, a growing number of organizations are interested in addressing this issue by leveraging the high-performance interconnects, such as Infini-
Band and 10 Gigabit Ethernet, which have been popular in High-Performance
Computing (HPC) Community. There is a lack of comprehensive examination of
the performance impact of these interconnects on MapReduce programs.

In this work, we systematically evaluate the performance impact of two popular high-speed interconnects, 10 Gigabit Ethernet and InfiniBand, using the
original Apache Hadoop and our extended Hadoop Acceleration framework. Our
analysis shows that, under the Apache Hadoop, although using fast networks can
efficiently accelerate the jobs with small intermediate data sizes, it is unable
to maintain such advantages for jobs with large intermediate data. In contrast,
Hadoop Acceleration provides better performance for jobs of a wide range of
data sizes. In addition, both implementations exhibit good scalability under different networks. Hadoop Acceleration significantly reduces CPU utilization and
I/O wait time of MapReduce programs.

1 Introduction

MapReduce, introduced by Google, has evolved as the backbone framework for
massive-scale data analysis. Its simple yet expressive interfaces, efficient scalability,
and strong fault-tolerance have attracted a growing number of organizations to build
their cloud services on top of the MapReduce framework. Hadoop MapReduce [1], advanced by Apache foundation, is a popular open source implementation of MapReduce
programming model. Compliant with MapReduce framewrok, Hadoop divides a job
into two types of tasks, called MapTasks and ReduceTasks, and assigns them to different machines for parallel processing. Although this framework is straightforward, its

---

* This research is supported in part by an NSF grant #CNS-1059376, and a grant from Lawrence
  Livermore National Laboratory.

T. Rabl et al. (Eds.): WBDB 2012, LNCS 8163, pp. 148–163, 2014.

intermediate data shuffling remains a critical and time-consuming operation as identified by many previous works [2, 3, 4, 5, 6]. Such data shuffling stage requires to move all the intermediate data generated by MapTasks to ReduceTasks, thereby causing a significant volume of network traffics and constraining the efficiency of data analytics applications.

High-Performance interconnects, such as InfiniBand [7] and 10 Gigabit Ethernet, provide appealing solutions to relieve such pressure on the intermediate data shuffling in the MapReduce frameworks. The state-of-the-art high-speed networks, such as InfiniBand, have been able to deliver up to 56Gbps bandwidth and sub-microsecond latency. Their low CPU utilization ability can spare more CPU cycles for MapReduce applications to accelerate their progress. Moreover, many fast interconnects have also offered Remote Direct Memory Access (RDMA) [8] protocol to fully take advantages of high-performance properties of those networks. Although high-performance interconnects have been popular in the High-Performance Computing (HPC) community, the performance impact of these networks on MapReduce programs remains unclear. A growing number of cloud companies, such as Amazon [9] are planning to build their next generation clusters on top of high-performance interconnects. But there is a lack of comprehensive examination of the performance impact of these interconnects on MapReduce programs.

To fill this void, we conduct a thorough assessment of the performance of MapReduce programs on different high-performance interconnects. In particular, we investigate how InfiniBand and 10 Gigabit Ethernet (10 GigE) improve the performance of the original Hadoop and our extended Hadoop Acceleration implementation [10]. Our evaluation centers around three aspects of MapReduce clusters, including scalability, performance impact on different phases, and resource utilization. Overall, our contributions can be summarized as the following:

- For the original Apache Hadoop, using high-performance interconnects can efficiently accelerate the programs with small intermediate data size by as much as 51.5%, but provide imperceptible performance improvement for programs with large intermediate data, when compared to 1 Gigabit Ethernet (1 GigE).
- The original Hadoop exhibits good scalability under 1/10 Gigabit Ethernets and InfiniBand environments. Compared to 1 GigE, 10 GigE and InfiniBand provide better scalability in large clusters.
- Simply adopting high-performance interconnects, the original Hadoop cannot reduce the CPU utilization and remove the disk bottleneck issues in existing design of Hadoop MapReduce.
- Our Hadoop Acceleration framework exhibits comparable scalability as the Apache Hadoop. Meanwhile it is able to efficiently speed up the jobs by as much as 49.5% in both InfiniBand and 10 GigE environments. Results also show that it accelerates both the map and reduce phases of data-intensive programs.
- Our Hadoop Acceleration cuts down on CPU utilization by up to 46.4% and alleviates disk contention by leveraging the advantages of fast networks.

The remainder of the paper is organized as follows. We briefly introduce the background in Section 2. We then present the comprehensive assessment results in Section 3.

Finally, we provide a review of related work in Section 4 and then conclude the paper in Section 5.

2 Background

2.1 Architecture of Apache Hadoop MapReduce

Hadoop implements MapReduce framework with two categories of components: a Job-Tracker and many TaskTrackers. The JobTracker commands TaskTrackers to process data in parallel through two main functions: map and reduce. In this process, the Job-Tracker is in charge of scheduling map tasks (MapTasks) and reduce tasks (Reduce-Tasks) to TaskTrackers. It also monitors their progress, collects run-time execution statistics, and handles possible faults and errors through task re-execution.

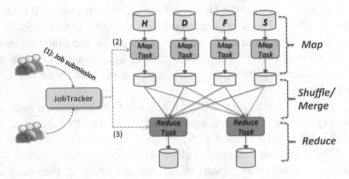

Fig. 1. An Overview of Data Processing in Hadoop MapReduce Framework

From the view of pipelined data processing, Hadoop consists of three main execution phases: map, shuffle/merge, and reduce as shown in Figure 1. In the first phase, the JobTracker selects a number of TaskTrackers and schedules them to run the map function. Each TaskTracker launches several MapTasks, one per split of data. The mapping function in a MapTask converts the original records into intermediate results, which are data records in the form of <key',val'> pairs. These new data records are stored as a MOF (Map Output File), one for every split of data. In the second phase, when MOFs are available, the JobTracker selects a set of TaskTrackers to run the ReduceTasks. TaskTrackers can spawn several concurrent ReduceTasks. Each ReduceTask starts by fetching a partition that is intended for it from a MOF (also called segment). Typically, there is one segment in each MOF for every ReduceTask. So, a ReduceTask needs to fetch such segments from all MOFs. Globally, these fetch operations lead to an all-to-all **shuffle** of data segments among all the ReduceTasks. This stage is also commonly referred as the **shuffle/merge** phase. In the third, or **reduce** phase, each ReduceTask loads and processes the merged segments using the reduce function. The final result is then stored to Hadoop Distributed File System [11].

Although the framework of Hadoop MapReduce is simple, the global all-to-all data shuffling process imposes significant pressure on the network. Therefore, it is appealing to accelerate such intermediate data shuffling via leveraging high-performance interconnects, such as InfiniBand or 10 Gigabit Ethernet (10 GigE). While, due to the higher cost of fast networks than traditional 1 Gigabit Ethernet (1 GigE), it is critical to investigate the benefits for Hadoop MapReduce to employ high performance networks.

2.2 Architecture of Hadoop-Acceleration

In order to address the issues exposed by intermediate data shuffling in Apache Hadoop, a new framework, called Hadoop Acceleration (Hadoop-A) [10], was designed on top of Apache Hadoop to take advantage of high-speed interconnects to accelerate the data movement.

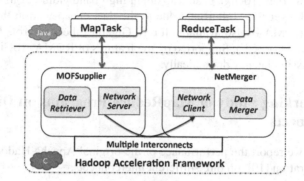

Fig. 2. Software Architecture of Hadoop-A

Figure 2 depicts the architecture of Hadoop Acceleration framework. Two new user-configurable plugin components, *MOFSupplier* and *NetMerger*, are introduced to leverage different protocols provided by the fast networks and enable alternative data merge algorithms. So far, the transportation protocols supported by Hadoop-A include TCP/IP and RDMA verbs. User have options to choose either Network Levitated Merge [10] or Hierarchical Merge algorithm [12] to conduct the merge process at the ReduceTask sides. Both algorithms strive to fully utilize the memory to accomplish the merge process, thus avoiding the disk I/O overhead. Both MOFSupplier and NetMerger are threaded C implementations, with all components following the object-oriented principle. We briefly describe several features of this acceleration framework without going too much into the technical details of their implementations.

User-Transparent Plugins – A primary requirement of Hadoop-A is to maintain the same programming and control interfaces for users. To this end, MOFSupplier and NetMerger plugins are designed as C processes that can be launched by TaskTrackers.

Multithreaded and Componentized MOFSupplier and Netmerger – MOFSupplier contains an network server that handles fetch requests from ReduceTasks. It also contains a data engine that manages the index and data files for all MOFs that are generated by local MapTasks. Both components are implemented with multiple threads in MOFSupplier. NetMerger is also a multithreaded program. It provides one thread for each Java ReduceTask. It also contains other threads, including a network client that fetches data partitions and a staging thread that uploads data to the Java-side ReduceTask.

2.3   Overview of High-Performance Interconnects

InfiniBand is a highly scalable interconnect technology, which can achieve low latency and high bandwidth that is up to 56Gbps. It is widely used in large data center, high performance computing systems and embedded applications, which require high speed communications. Featured by Remote Direct Memory Access (RDMA), InfiniBand transfers data directly between memory locations over network without the involvement of CPU and extra data copying. And because InfiniBand incurs very low CPU utilization, it is ideal to carry several traffic categories, like management data and storage data, over a single connection.

10 Gigabit Ethernet (10GigE) can also attain high bandwidth but its data transmission latency is longer than InfiniBand due to the data encapsulation through TCP/IP protocol stack. RDMA is also available for 10 Gigabit Ethernet through RDMA over Converged Ethernet (RoCE). Supported by the features of RDMA, 10GigE can reduce the data transmission latency dramatically.

3   Benchmarking Study of MapReduce Programs on Different Interconnects

In this section, we report the performance of the original Apache Hadoop [1] and our Hadoop Acceleration [10], on three different interconnects.

3.1   Experimental Environment

All experiments are conducted on two environments, which are InfiniBand environment and Ethernets environment, respectively. Each environment features 23 compute nodes. All compute nodes in both clusters are identical. Each node is equipped with four 2.67GHz hex-core Intel Xeon X5650 CPUs, two Western Digital SATA 7200 RPM 500GB hard drives and 24GB memory.

In the Ethernet environment, all compute nodes connect to both a 1 Gigabit NET-GEAR switch and a 10 Gigabit Voltaire switch. In the InfiniBand environment, Mellanox ConnectX-2 QDR Host Channel Adaptors are installed on each node that connect to a 108-port InfiniBand QDR switch providing up to 40 Gb/s full bisection bandwidth per port. We use the InfiniBand software stack, OpenFabrics Enterprise Distribution (OFED) [13] version 1.5.3, as released by Mellanox. Note that in the InfiniBand environment, IPoIB (an emulated implementation of TCP/IP on InfiniBand) provides standardized IP encapsulation over InfiniBand links. Therefore, all applications that require TCP/IP can continue to run without any modification. Detailed description of IPoIB can be found in [14]. InfiniBand also provides Socket Direct Protocol (SDP) [15] to accelerate the TCP/IP protocol via leveraging the RDMA capability, albeit its performance is still not as competive as the RDMA verbs. Similar to IPoIB, it requires no modification to the applications. Recently, OFED has announced to discontinue the support for SDP. However, we still include the evaluation results with SDP in this work to provide insight for its potential successor protocol, such as jVerbs. Although both InfiniBand and 10 GigE provide RDMA protocol, current Hadoop is unable to directly use it but via the SDP protocol.

In terms of the Hadoop setup, we employ the stable version Hadoop 1.0.4. During the experiments, one node is dedicated for both the NameNode of HDFS and the JobTracker of Hadoop MapReduce. On each slave nodes, we allocate 4 MapTask and 2 ReduceTask slots. The HDFS block size is chosen as 256MB as suggested by [16] to balance the parallelism and performance of MapTasks. We assign 512 MB and 1 GB heap size to each MapTask and ReduceTask respectively.

Benchmarks we adopt to conduct the evaluation include Terasort, WordCount, from standard Hadoop package. Terasort is extensively used through the entire evaluation due to its popularity as a *de facto* standard Hadoop I/O benchmark. In the Terasort, the size of intermediate data and the final output are as large as the input size. Via controlling the input data size, Terasort can effectively expose the I/O bottleneck across the Hadoop data processing pipeline.

Many cases have been explored in our evaluation experiments. To avoid confusion, we list the protocol and network environment used for each test case in Table 1. In addition, in the following sections, we use Hadoop-A and Hadoop Acceleration interchangeably.

Table 1. Protocol and Network Description of Test Cases

Name of Test Cases	Transport Protocol	Network
Apache Hadoop with 1GigE	TCP/IP	1 GigE
Apache Hadoop with 10GigE	TCP/IP	10 GigE
Apache Hadoop with IPoIB	IPoIB	InfiniBand
Apache Hadoop with SDP	SDP	InfiniBand
Hadoop-A with 10 GigE	TCP/IP	10 GigE
Hadoop-A with RoCE	RoCE	10 GigE
Hadoop-A with RDMA	RDMA	InfiniBand
Hadoop-A with IPoIB	IPoIB	InfiniBand

3.2   Impact of High-Performance Interconnects on Hadoop MapReduce

We have evaluated the impact of high-performance interconnects on Apache Hadoop MapReduce from three aspects, which are *scalability, performance impact on different phases*, and *resource utilization*.

Scalability. We study the scalability of Apache Hadoop via examining its ability to process a growing amount of data with fixed amount of computational resources and its ability to improve the throughput when expending the resources (a.k.a *Scale Up*).

To investigate the ability of Hadoop to process growing amount of data, we run Terasort jobs of different input sizes on 22 slave nodes in both InfiniBand and Ethernet environments. For each data size, we conduct 3 experiments and report the average job execution time. Overall, Hadoop shows linear scalability for small data sizes and nonlinear increase for large data sets.

Figure 3 shows the performance of Apache Hadoop with various data sizes under 3 different networks. As shown in Figure 3 (a), compared to running Hadoop on 1 GigE, using 10 GigE reduces the job execution time by 26.5% on average. Noticeably, fast

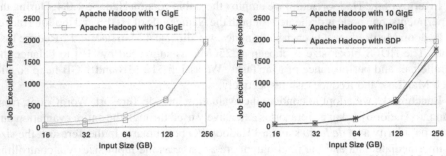

(a) Comparison between 1 GigE and 10 GigE   (b) Comparison between 10 GigE and Infini-
Band

Fig. 3. Performance of Apache Hadoop with Growing Data Sizes

network is very beneficial for small data sizes ($\leq$ 64 GB). For instance, when the data
size is 32GB, compared to Hadoop on 1GigE, using 10 GigE effectively speeds up
the job execution time by as much as 51.5%. This is because the movement of small
size data is less dependent on disks and most of them reside in disk cache or system
buffers. Thus high-performance networks can exhibit better benefits for data shuffling.
While, simply adopting fast networks provide no encouraging improvements for large
data sets ($\geq$ 128 GB) due to severe disk I/O bottleneck caused by large data sets. Such
disk bottleneck is triggered in many places across the data shuffling. In particular, when
the merge process is heavy, a large number of small random reads for retrieving the
merged results exist in ReduceTasks and quickly vanish the improvements gained from
fast data movement.

On the other hand, as shown in the Figure 3 (b), although InfiniBand provides even
higher throughput and lower latency than 10 GigE, using InfiniBand achieves negligible
performance improvements across the tests and only slightly reduces the job execution
time by 13% for 256 GB data size due to less memory copy overhead involved in IPoIB
and SDP.

We further study the Apache Hadoop's ability to scale up with two patterns, which
are *Strong Scaling* pattern and *Weak Scaling* pattern. In the case of strong scaling, we
fix the input data size (256GB) while expending the number of slave nodes. In the
case of weak scaling, for each test case, we use a fixed-size data size (6GB) for each
ReduceTask, so the total input size increases linearly when we increase the number of
nodes, reaching 264GB when 22 slave nodes are used.

Figure 4 shows the results of strong scaling tests under different networks. In all of
the test cases, job execution time reduces linearly as more nodes join the computation.
However, as shown in the Figure 4 (a), on average, 10 GigE only marginally accelerates the job execution by 8.1% due to disk constraints. In addition, we observe that in
the strong scaling case, 10 GigE even delivers better performance than InfiniBand. On
average, 10 GigE outperforms IPoIB and SDP by 3.2% and 5.8%. Such results demonstrate that InfiniBand do not have superior advantages over 10 GigE for data-intensive
MapReduce applications.

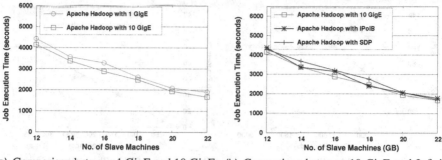

(a) Comparison between 1 GigE and 10 GigE   (b) Comparison between 10 GigE and Infini-
Band

Fig. 4. Strong Scaling Evaluation of Apache Hadoop

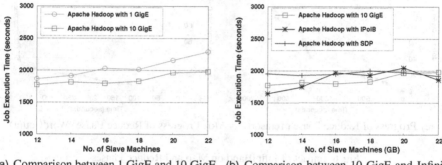

(a) Comparison between 1 GigE and 10 GigE   (b) Comparison between 10 GigE and Infini-
Band

Fig. 5. Weak Scaling Evaluation of Apache Hadoop

For the weak scaling tests, the optimal result should be a uniform execution time across the tests. However, as shown in Figure 5, in the 1 GigE environment, job is slowed down by 22.8% when the number of nodes increases from 12 to 22, showing poor scalability of 1 GigE. In contrast, using 10 GigE not only decreases the job execution time by 21.9% on average, but achieves better scalability (11.6% increase from 12 to 22 nodes) as well. Similar to the strong scaling tests, 10 GigE outperforms Infini-Band in the weak scaling test with respect to the job execution time. This is because when the number of nodes is small, 10 GigEs TCP/IP protocol is more lightweight than IPoIB protocol, which is an emulation of TCP/IP protocol in the InfiniBand environment, leading to more overhead. But InfiniBand shows better scalability for large cluster size due to higher bandwidth and the design of InfiniBand HCA. When the number of nodes increases from 12 to 22, job execution is slightly degraded by 1.7% when running with SDP protocol.

Impact on Different Phases of the Data Processing Pipeline. To evaluate the impact of high-performance interconnects on different map and reduce phases of different types of Apache Hadoop applications. We employ two applications, Terasort and

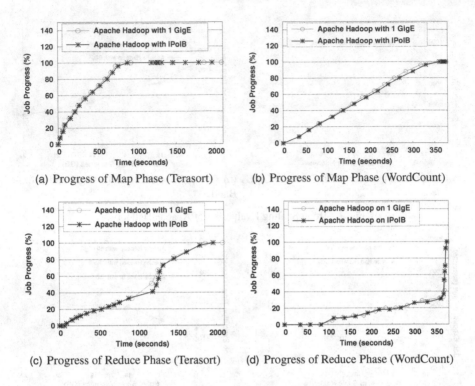

(a) Progress of Map Phase (Terasort)   (b) Progress of Map Phase (WordCount)

(c) Progress of Reduce Phase (Terasort)   (d) Progress of Reduce Phase (WordCount)

Fig. 6. Impact of InfiniBand on Map and Reduce Phases

WordCount, which represent data-intensive and computation-intensive application, respectively. Figure 6 depicts the progress of map and reduce phases of different jobs.

Figure 6 (a) and (b) show that simply running Apache Hadoop on InfiniBand gain imperceptible performance improvement in both applications. In Terasort job, each MapTask spends a large portion of task execution time on merging the temporary spilled files. While the MapTasks in WordCount consume more CPU resources. Thus both types of MapTasks provide InfiniBand with limited optimization spaces. While on the ReduceTask side, as shown in Figure 6 (c) and (d), we observe that InfiniBand still fails to accelerate the progress of ReduceTasks of Terasort due to extremely slow merge process within the ReduceTasks. Expensive merge process and repetitive merge behavior significantly drag down the draining of data from the network links, resulting in slow ReduceTasks. For the WordCount, InfiniBand offers negligible improvement due to very small intermediate data sizes. Since 10 GigE exibits similar performance pattern, we omit to elaborate on its results here for conciseness.

Resource Utilization. In addition to job execution times, we have also collected the CPU and disk utilization statistics across the experiments. CPU utilization is an important performance metric. Low CPU utilization during data shuffling and merging can spare more CPU cycles for acceleration computation phases of Hadoop applications. Figure 7 (a) shows the CPU utilization of Apache Hadoop when running Terasort

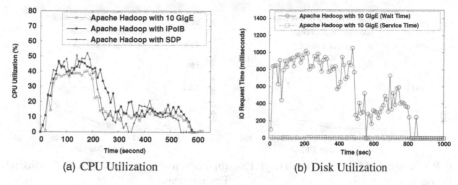

(a) CPU Utilization   (b) Disk Utilization

Fig. 7. Analysis of CPU and Disk Utilization of Apache Hadoop

benchmark with 128GB input data. As shown in this figure, utilization difference between 10 GigE and InfiniBand is not significant. During the first 200 seconds, 10 GigE presents slightly lower utilization. The reason is that in the InfiniBand tests cases, there are more remote MapTasks, meaning more MapTasks need to fetch input data remotely. After 200 seconds, three networks achieve very similar utilization statistics. Compared to the 1 GigE, three networks reduces the CPU utilization by 9.6% on average (for clear presentation, we omit the 1 GigE data in Figure 7 (a)).

As we have claimed before, disk bound merge process is the major performance bottleneck that prevents Apache Hadoop from making full use of advantages of fast networks. To illustrate such issue, we have examined the I/O wait (queuing) time and the service time of I/O requests during a 160 Terasort job execution. As shown in Figure 7 (b), the I/O wait time can be more than 1000 milliseconds. Worse yet, extremely quick service time (< 8 milliseconds) indicates that most I/O requests are spending nearly 100% of their time waiting in the queue , which explicitly demonstrates that the storage system is not able to keep up with the requests. In addition, both MapTasks and ReduceTasks intensively competes for disk bandwidth, this can significantly overload the disk subsystem, causing high-performance networks unable to accelerate the data-intensive MapReduce applications.

3.3   Impact of High-Performance Interconnects on Hadoop Acceleration

Hadoop Acceleration framework is a major step forward for Hadoop to support high-performance networks. Its network layer is designed to be highly portable on a variety of networks, such as InfiniBand and 10 GigE with a rich set of protocol support, such as RDMA, TCP/IP, etc. In this section, we study the performance of Hadoop Acceleration framework (Hadoop-A) on different networks via the same strategies used in section 3.2 and compare its performance against original Apache Hadoop.

Scalability. Figure 8 shows the performance of Hadoop-A with growing data sets by conducting Terasort benchmark. Overall, Hadoop-A is superior to the original Apache Hadoop for data-intensive applications. In the 10 GigE environment, compared to Apache Hadoop, running Hadoop-A with TCP/IP reduces the job execution times by

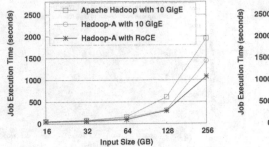
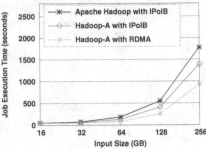

(a) Performance of Hadoop-A in 10 GigE Environment
(b) Performance of Hadoop-A in InfiniBand Environment

Fig. 8. Performance of Hadoop Acceleration with Growing Data Sizes

19.3% on average. Enabling the RDMA over Converged Ethernet (RoCE) protocol further accelerates the job execution by up to 15.3%. In the InfiniBand environment, on average, Hadoop-A outperforms the original Hadoop by 14.1% and 38.7%, when IPoIB and RDMA protocols are used respectively. Moreover, in both environments, we observe that Hadoop-A delivers better performance for all ranges of data sizes. For small data sets, Hadoop-A is better due to its elimination of JVM overhead from the critical path of data shuffling. For large data sets, Hadoop-A mitigates the disk bottleneck via its flexible framework to support various merging algorithms, such as Network Levitated Merge [10] and Hierarchical Merge algorithm [12], both of which fully take advantage of the memory to avoid disk I/O for the intermediate data merging process at the ReduceTask side.

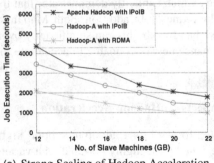
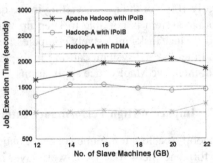

(a) Strong Scaling of Hadoop Acceleration
(b) Weak Scaling of Hadoop Acceleration

Fig. 9. Scalability Evaluation of Hadoop Acceleration in InfiniBand Environment

Scalability is a critical feature for MapReduce, thus we also evaluate the strong scaling and weak scaling capability of Hadoop-A. As shown in the Figure 9 (a), running Hadoop-A with RDMA and with IPoIB effectively outperform Apache Hadoop by 49.5% and 20.9%, respectively on average, and achieves a comparable linear reduction as the original Hadoop. On the other hand, for the weak scaling tests, Hadoop-A with RDMA and with IPoIB reduce the execution time by 43.6% and 21.1%, respectively on average, compared to Apache Hadoop with IPoIB, and exhibit slight performance variance across the tests. (Similar performance is observed under the 10 GigE environment, so we omit the results for succinctness).

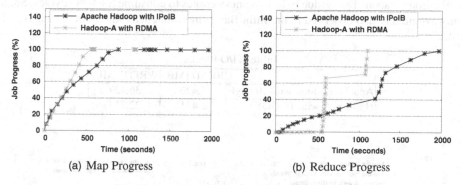

(a) Map Progress   (b) Reduce Progress

Fig. 10. Impact on Different Phases of Terasort

Impact on Different Phases of Data Processing Pipeline. Across the tests, we observe that Hadoop-A is able to speedup both the map and reduce phases of the data-intensive applications. Figure 10 illustrates such phenomenon, in which MapTasks of TeraSort complete much faster under Hadoop-A, especially when the percentage of completion goes over 50%. This is because in Hadoop-A, all ReduceTasks only performs lightweight operations such as fetching headers and setting up in-memory priority queue during the map phase, thereby leaving more resources such as disk bandwidth for MapTasks, which can fully relish the resources to accelerate their executions. While, on the ReduceTask side, we observe slow progress at the beginning (before 500th second). This is because during this period, ReduceTasks are just constructing the in-memory priority queue, meanwhile waiting for the map phase to complete. As soon as the map phase complete (after 500th second), ReduceTask rapidly progresses to the 60%, significantly outperforming the original ReduceTasks.

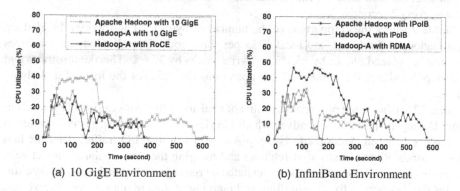

(a) 10 GigE Environment   (b) InfiniBand Environment

Fig. 11. CPU Utilization of Hadoop Acceleration

Resource Utilization. By eliminating the overhead of JVM and reducing the disk I/O, Hadoop-A greatly lowers the CPU utilization. As shown in the Figure 11 (a), in the 10

GigE environment, compared to Apache Hadoop on 10GigE, Hadoop-A on RoCE and on 10GigE reduce the CPU utilization by 46.4% and 33.9%, respectively on average. In addition, compared to TCP/IP protocol, leveraging RoCE cuts down on the CPU utilization by about 18.7% due to less memory copies to consume the CPU cycles. Such improvement is also observed in the InfiniBand environment as shown in the Figure 11 (b).

Table 2. I/O Blocks

	READ (MB)	WRITE (MB)
Apache Hadoop with 10 GigE	5,426	36,427
Hadoop-A with 10 GigE	2,441	22,713

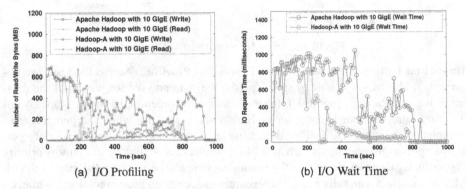

(a) I/O Profiling   (b) I/O Wait Time

Fig. 12. Disk Utilization of Hadoop Acceleration

To assess the effectiveness of Hadoop-A to improve the disk utilization, we have also measured the disk accesses during data shuffling under Hadoop-A, and compared the results with that of Apache Hadoop. We run TeraSort on 20 slave nodes with 160GB as input size. On each node we run vmstat and iostat to collect I/O statistics and trace the output every 2 seconds.

Table 2 shows the comparison of the number of bytes read and written by Hadoop and Hadoop-A from and into local disks per slave node. Overall, Hadoop-A reduces the number of read blocks by 55.1% and write blocks by 37.6%. This demonstrates that Hadoop-A reduces the number of I/O operations and relieves the load of underlying disks.

Fig. 12 (a) shows the progressive profile of read and write bytes during the job execution. During the first 200 seconds in which MapTasks are active, there is no substantial difference between Hadoop and Hadoop-A in terms of disk I/O traffic. After the first 200 seconds, ReduceTasks start fetching and merging the intermediate data actively. Because Hadoop-A uses the network-levitated merge algorithm which completely eliminates the disk access for the shuffling and merging of data segments, we observe that Hadoop-A effectively reduces the number of bytes read from or written to the disks. Therefore, disk I/O traffic is significantly reduced during this period.

As shown in section 3.2, I/O requests in Apache Hadoop experience long I/O wait time, thus degrading the performance. In order to further analyze the benefit from the reduced disk accesses, we measure the I/O wait time in Hadoop-A, and compare it with Apache Hadoop. The result is shown in Figure 12 (b). As shown in the figure, Hadoop-A leads to similar or lower I/O wait time during the first 200 seconds, which corresponds to the mapping phase of the job execution. As the job progresses, I/O wait time of Hadoop-A is significantly reduced when job enters into the shuffle/merge and reduce phases. This demonstrates that the reduction of disk accesses contributes to the reduction of I/O wait time. Aggregately, these experiments indicate that Hadoop-A effectively improves I/O performance in Hadoop, thereby effectively shortening job execution time.

4 Related Work

Leveraging high performance interconnects to move data in the Hadoop ecosystem has attracted numerous research interests from many organizations over the past a few years. Huang et al. [17] designed an RDMA-based HBase over InfiniBand. In addition, they pointed out the disadvantages of using Java Socket Interfaces in Hadoop ecosystem. A recent evaluation [18] of Hadoop Distributed File system (HDFS) [11] used the SDP [15] and IPoIB protocols of InfiniBand [19] to investigate the potential benefit of leveraging fast networks for pipelined writing in HDFS. In the same work, authors showed that Hadoop was unable to directly leverage the RDMA (Remote Direct Memory Access) communication mechanism available from high-performance RDMA interconnects. For that reason, to enhance the efficiency of HDFS, Islam et al. [20] modified HDFS network connection structure to use RDMA over InfiniBand via JNI interfaces. Jose et al. [21, 22] implemented a scalable memcached through taking advantage of performance benefits provided by InfiniBand. But, although Hadoop MapReduce is a fundamental basis of Hadoop ecosystem, there is lack of research on how to efficiently leverage high performance interconnects in Hadoop MapReduce. [10] studied the feasibility of importing RDMA support into the Apache Hadoop. Meanwhile, [23] measured the overhead imposed by Java Virtual Machine on the Hadoop shuffling.

Adopting RDMA from high speed networks for fast data movement has been very popular in various programming models and storage paradigms, such as MPI. [24] studied the pros and cons of using RDMA capabilities in a great details. Liu et al. [25] designed RDMA-based MPI over InfiniBand. Yu et al. [26] implemented a scalable connection management strategy for high-performance interconnects. Implementations of PVFS [27] on top of RDMA networks such as InfiniBand and Quadrics were described in [28] and [29], respectively. However, none of them have studied the impact of high-speed inter-connection on Hadoop MapReduce framework.

5 Conclusions

In the Hadoop MapReduce framework, data shuffling accounts for a considerable portion of the total execution time of MapReduce programs. Meanwhile, the current technologies in interconnect fabric have made the speed of NIC comparable with RAM-base memory. In this paper, we undertake a comprehensive evaluation on how Hadoop

MapReduce framework can be accelerated by leveraging high-performance interconnects. We have examined the performance of Apache Hadoop and Hadoop Acceleration framework on InfiniBand and 1/10 Gigabit Ethernet from the aspects of scalability, data processing pipeline and resource utilization. Our experiment results reveal that simply switching to the high-performance interconnects cannot effectively boost the performance of Apache Hadoop due to the cost imposed by JVM and disk bottleneck on Hadoop intermediate data shuffling. Moreover, with various application evaluated on both Ethernet and InfiniBand environments, we demonstrate that Hadoop Acceleration framework can significantly reduce the CPU utilization and job execution time for MapReduce jobs that generate a large amount of intermediate data. Specifically, Hadoop Acceleration can effectively reduce the execution time of Hadoop jobs by up to 49.5% and lower the CPU utilization by 46.4%. In the future, we plan to further evaluate the MapReduce programs on large clusters that consists of hundreds of thousands nodes.

References

[1] Apache Hadoop Project, http://hadoop.apache.org/
[2] Dean, J., Ghemawat, S.: Mapreduce: simplified data processing on large clusters. In: Proceedings of the 6th Conference on Symposium on Opearting Systems Design & Implementation, OSDI 2004, vol. 6, p. 10. USENIX Association, Berkeley (2004)
[3] Pavlo, A., Paulson, E., Rasin, A., Abadi, D.J., DeWitt, D.J., Madden, S., Stonebraker, M.: A comparison of approaches to large-scale data analysis. In: Proceedings of the 35th SIGMOD International Conference on Management of Data, SIGMOD 2009, pp. 165–178. ACM, New York (2009)
[4] Condie, T., Conway, N., Alvaro, P., Hellerstein, J.M., Elmeleegy, K., Sears, R.: Mapreduce online. In: Proceedings of the 7th USENIX Conference on Networked Systems Design and Implementation, NSDI 2010, p. 21. USENIX Association, Berkeley (2010)
[5] Seo, S., Jang, I., Woo, K., Kim, I., Kim, J.S., Maeng, S.: HPMR: Prefetching and pre-shuffling in shared MapReduce computation environment. In: CLUSTER, pp. 1–8 (August 2009)
[6] Rao, S., Ramakrishnan, R., Silberstein, A., Ovsiannikov, M., Reeves, D.: Sailfish: a framework for large scale data processing. In: Proceedings of the Third ACM Symposium on Cloud Computing, SoCC 2012, pp. 4:1–4:14. ACM, New York (2012)
[7] InfiniBand Trade Association: The InfiniBand Architecture, http://www.infinibandta.org
[8] Recio, R., Culley, P., Garcia, D., Hilland, J.: An rdma protocol specification, version 1.0 (October 2002)
[9] High Performance Computing (HPC) on AWS, http://aws.amazon.com/hpc-applications/
[10] Wang, Y., Que, X., Yu, W., Goldenberg, D., Sehgal, D.: Hadoop acceleration through network levitated merge. In: Proceedings of 2011 International Conference for High Performance Computing, Networking, Storage and Analysis, SC 2011, pp. 57:1–57:10. ACM, New York (2011)
[11] Shvachko, K., Kuang, H., Radia, S., Chansler, R.: The hadoop distributed file system. In: Proceedings of the 2010 IEEE 26th Symposium on Mass Storage Systems and Technologies, MSST 2010, pp. 1–10. IEEE Computer Society, Washington, DC (2010)

[12] Que, X., Wang, Y., Xu, C., Yu, W.: Hierarchical merge for scalable mapreduce. In: Proceedings of the 2012 Workshop on Management of Big Data Systems, MBDS 2012, pp. 1–6. ACM, New York (2012)

[13] Open Fabrics Alliance, http://www.openfabrics.org

[14] Chu, J., Kashyap, V.: Transmission of IP over InfiniBand(IPoIB) (2006), http://tools.ietf.org/html/rfc4391

[15] InfiniBand Trade Association: Socket Direct Protocol Specification V1.0 (2002)

[16] Zaharia, M., Borthakur, D., Sen Sarma, J., Elmeleegy, K., Shenker, S., Stoica, I.: Delay scheduling: a simple technique for achieving locality and fairness in cluster scheduling. In: Proceedings of the 5th European Conference on Computer Systems, EuroSys 2010, pp. 265–278. ACM, New York (2010)

[17] Huang, J., Ouyang, X., Jose, J., Wasi-ur-Rahman, M., Wang, H., Luo, M., Subramoni, H., Murthy, C., Panda, D.K.: High-performance design of hbase with rdma over infiniband. In: 26th IEEE International Parallel and Distributed Processing Symposium, IPDPS 2012, Shanghai, China, May 21-25, pp. 774–785 (2012)

[18] Sur, S., Wang, H., Huang, J., Ouyang, X., Panda, D.K.: Can High-Performance Interconnects Benefit Hadoop Distributed File System? In: MASVDC 2010 Workshop in Conjunction with MICRO (December 2010)

[19] Infiniband Trade Association, http://www.infinibandta.org

[20] Islam, N.S., Rahman, M.W., Jose, J., Rajachandrasekar, R., Wang, H., Subramoni, H., Murthy, C., Panda, D.K.: High performance rdma-based design of hdfs over infiniband. In: Proceedings of 2012 International Conference for High Performance Computing, Networking, Storage and Analysis, SC 2012. ACM (2012)

[21] Jose, J., Subramoni, H., Luo, M., Zhang, M., Huang, J., Wasi-ur-Rahman, M., Islam, N.S., Ouyang, X., Wang, H., Sur, S., Panda, D.K.: Memcached design on high performance rdma capable interconnects. In: ICPP, pp. 743–752. IEEE (2011)

[22] Jose, J., Subramoni, H., Kandalla, K., Wasi-ur Rahman, M., Wang, H., Narravula, S., Panda, D.K.: Scalable memcached design for infiniband clusters using hybrid transports. In: Proceedings of the 2012 12th IEEE/ACM International Symposium on Cluster, Cloud and Grid Computing (CCGrid 2012), pp. 236–243. IEEE Computer Society, Washington, DC (2012)

[23] Wang, Y., Xu, C., Li, X., Yu, W.: Jvm-bypass for efficient hadoop shuffling. In: 27th IEEE International Parallel and Distributed Processing Symposium, IPDPS 2013. IEEE (2013)

[24] Frey, P.W., Alonso, G.: Minimizing the hidden cost of rdma. In: Proceedings of the 2009 29th IEEE International Conference on Distributed Computing Systems, ICDCS 2009, pp. 553–560. IEEE Computer Society, Washington, DC (2009)

[25] Liu, J., Wu, J., Panda, D.K.: High performance rdma-based mpi implementation over infiniband. International Journal of Parallel Programming 32, 167–198 (2004)

[26] Yu, W., Gao, Q., Panda, D.K.: Adaptive connection management for scalable mpi over infiniband. In: Proceedings of the 20th International Conference on Parallel and Distributed Processing, IPDPS 2006, p. 102. IEEE Computer Society, Washington, DC (2006)

[27] Carns, P.H., Ligon III, W.B., Ross, R.B., Thakur, R.: PVFS: A Parallel File System For Linux Clusters. In: Proceedings of the 4th Annual Linux Showcase and Conference, Atlanta, GA, pp. 317–327 (October 2000)

[28] Wu, J., Wychoff, P., Panda, D.K.: PVFS over InfiniBand: Design and Performance Evaluation. In: Proceedings of the International Conference on Parallel Processing 2003, Kaohsiung, Taiwan (October 2003)

[29] Yu, W., Liang, S., Panda, D.K.: High Performance Support of Parallel Virtual File System (PVFS2) over Quadrics. In: Proceedings of the 19th ACM International Conference on Supercomputing, Boston, Massachusetts (June 2005)

BigBench Specification V0.1
BigBench: An Industry Standard Benchmark for Big Data Analytics

Tilmann Rabl[1], Ahmad Ghazal[2], Minqing Hu[2], Alain Crolotte[2],
Francois Raab[3], Meikel Poess[4], and Hans-Arno Jacobsen[1]

[1] University of Toronto
[2] Teradata Corp.
[3] InfoSizing Inc.
[4] Oracle Corp.

Abstract. In this article, we present the specification of BigBench, an end-to-end big data benchmark proposal. BigBench models a retail product supplier. The benchmark proposal covers a data model and a set of big data specific queries. BigBench's synthetic data generator addresses the variety, velocity and volume aspects of big data workloads. The structured part of the BigBench data model is adopted from the TPC-DS benchmark. In addition, the structured schema is enriched with semi-structured and unstructured data components that are common in a retail product supplier environment. This specification contains the full query set as well as the data model.

1 Introduction

Big data (BD) is about increasing volume of data from a variety of sources including structured, semi-structured and unstructured data. Some of the BD sources are typically generated with high velocity like click streams and sensors logs. This wealth of data provides a lot of new analytic and business intelligence (BI) opportunities like fraud, churn and customer loyalty analysis.

Many commercial and open source systems were built or extended to store and process BD. These tools are mostly parallel database management systems or MapReduce (MR) based systems. There are no standards yet on BD processing, but for the most part these systems provide SQL, UDF, MR or a mix of these as an interface.

Even though there are no standards for BD yet, still there is a need to measure and compare the performance of the different systems that claim to support BD. Recently, there are quite a few efforts in the area of big data benchmarking (e.g. PigMix[1], GridMix[2], GraySort[3]). Some of these benchmarks are focused on one component of the system and others are focused on specific MR systems.

---

[1] PigMix – https://cwiki.apache.org/confluence/display/PIG/PigMix
[2] GridMix – http://hadoop.apache.org/docs/mapreduce/current/gridmix.html
[3] Sort Benchmark Home Page – http://sortbenchmark.org

T. Rabl et al. (Eds.): WBDB 2012, LNCS 8163, pp. 164–201, 2014.
© Springer-Verlag Berlin Heidelberg 2014

In this article, we present the specification of the end to end big data bench-mark BigBench. BigBench is based on a fictitious retailer who sells products to customers via physical and online stores. This specification completes our previous publication that covered details on the data model, synthetic data generator, workload description, and metrics [1]. The workload queries are specified in English and in Teradata Aster's SQL-MR syntax [2,3]. We introduce new metrics specific to BD data loading and workload execution. The feasibility of the proposal is shown by applying it on the Teradata Aster DBMS (TAD). This experiment involves generating 200 gigabyte of data and loading it into TAD. The workload is executed as a single stream of queries.

The rest of this article is structured as follows. In Section 2, we describe the BigBench data model. In Section 3, we give a short overview of the BigBench data generation. We describe the BigBench workload in Section 4. Section 5 shows the results of our proof-of-concept evaluation of BigBench on Teradata Aster. We conclude in Section 6. In Appendix A, we list all 30 BigBench queries and Appendix B contains the complete schema for BigBench.

2 Data Model

BD is not about volume only. Douglas Laney described the 3 Vs of BD referring to volume, velocity and variety [4]. Velocity is an important issue in BD since such data like clicks or sensor information are produced at an increasing rate. Also, data comes in different forms like structured relational tables, semi-structured key-value web clicks or unstructured social data text. Our data model has the volume, variety and velocity elements as described in the following.

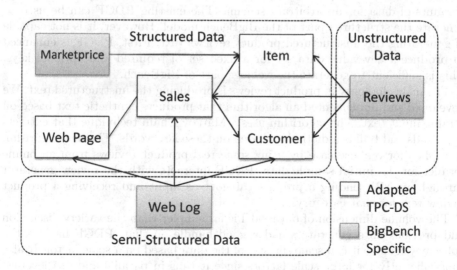

Fig. 1. Simplified BigBench Data Model

The variety property of our model is illustrated in Figure 1. The structured part of BigBench is an adaption of the TPC-DS model which also depicts a product retailer [5]. We borrowed the store and online sales part from that model and added a table for competitor prices of the retailer.

The structured part is enriched with semi-structured and unstructured data shown in the lower and right part of Figure 1. The semi-structured part's content is composed by clicks made by customers and guest users visiting the retailer site. Some of these clicks are for completing a customer order. As shown in Figure 1, the semi-structured data is logically related to the Web Page, Customer and Sales tables in the structured part. Our design assumes the semi-structured data to be a key-value format similar to Apache web server log format.

Typically, database and MR systems would convert such format to a table/file with a schema like (DateID, TimeID, SalesID, WebPageID, UserID). However, we do not require such conversion since some systems may choose to run analytics on the native key-value format. Product reviews is a growing source of online retail data. We found such source to be an excellent representation for the unstructured data in our model. Figure 1 shows product reviews in the right part and its relationship to Date, Time, Item, Users and Sales tables in the structured part. One implementation of the product reviews is a single table/file with a structure like (DateID, TimeID, SalesID, ItemID, ReviewRating, ReviewText). The full schema is specified in SQL in Apendix B.

3   Data Generation

Our work also provides a design and implementation of a data generator for the proposed BigBench data model. Our data generator is based on an extension of PDGF [6]. PDGF is a parallel data generator that is capable of producing large amounts of data for an arbitrary schema. The existing PDGF can be used to generate the structured part of the BigBench model. However, it is not capable of generating the unstructured product reviews text. First, PDGF is enhanced to produce a key-value data set for a fixed set of required and optional keys. This is sufficient to generate the weblogs part of BigBench.

The main challenge in product reviews is producing the unstructured text. We developed and implemented an algorithm that produces synthetic text based on sample input text. The algorithm uses a Markov Chain technique that extracts key words and builds a dictionary based on these key words. The new algorithm is applied for our use case by using some real product reviews from an online retailer for the initial sample data. PDGF interacts with the review generator through an API sending a product category as input and receiving a product review text for that category.

The volume dimension of our model is far simpler than the variety discussion and previous data generators had a good handle on that. PDGF handles the volume well since it can scale the size of the data based on a scale factor. It also runs efficiently for large scale factors since it runs in parallel and can leverage large systems dedicated for the benchmark.

For our proof-of-concept system, the tables that are originating from TPC-DS are generated using DSdgen, the TPC-DS standard data generator[4].

4   Workload

The second major component of BigBench is the specification of workload queries applied on the BigBench data model. In terms of business questions, we found the big data retail analytics by McKinsey serves our purpose given that BigBench is about retail [7]. In [7] five major areas of big data analytics are described namely: marketing, merchandising, operations, supply chain and new business models. These areas are further broken down into sub-functions. For example, marketing can be broken down into cross selling, sentiment analysis, etc. We used these 5 areas and added reporting as a sixth area. We postulate that a big data benchmark should have some traditional business intelligence or reporting type of queries.

In addition to the big data retail business levers above, we looked at the different technical aspects the BigBench queries should measure. We identified the following three areas:

- The type of the input data the query is addressing. We made sure each of the structured, semi-structured, unstructured and their combinations are covered in the queries. Out of the 30 queries 18 (60%) are exclusively on the structured data, 7 (23.3%) incorporated semi-structured data, and 5 (16.7%) additionally incorporated unstructured data.
- The type of processing appropriate for the query. This dimension targets the two common paradigms of SQL (and similar constructs like HQL) and MR. Thus, our queries can be answered by SQL, others by MR or a mix of both. Note that some of the perceived MR queries can also be written through complex SQL constructs like window functions and therefore we identify the two classes in this dimension as simple SQL and MR or complex SQL. In this definition, 12 (40%) queries are declarative (pure SQL), 5 (16.7%) queries are procedural (MR), and 13 (43.3%) are a mix.
- The third important technical dimension is the different algorithms of MR processing as described by the Apache MAHOUT system. Classes of algorithms used in the BigBench queries are statistical analysis (6 queries), path analysis (5 queries), text analysis (4 queries), association mining (4 queries), classification (1 query), clustering (3 queries), reporting (8 queries).

The categorization along technical dimensions with corresponding query numbers is shown in Table 4. The implementation technique is either declarative, procedural, or mixed. Declarative queries are pure SQL queries, that could also be processed by stock relational database systems. Procedural queries are pure MapReduce implementations that do not need joins. Mixed queries contain MapReduce functions along with relational operations, such as joins or views.

---

[4] TPC-DS and DSDgen is available at http://www.tpc.org/tpcds/default.asp

Query Type	Queries	Percent	Data Type	Queries	Percent
Declarative	6, 7, 9, 13, 14, 16, 17, 19, 21, 22, 23, 24	40%	Structured	1, 6, 7, 9, 13, 14, 15, 16, 17, 19, 20, 21, 22, 23, 24, 25, 26, 29	60%
Mixed	1, 4, 5, 8, 11, 12, 15, 18, 20, 25, 26, 29, 30	43%	Semi-Structured	2, 3, 4, 5, 8, 12, 30	23%
Procedural	2, 3, 10, 27, 28	17%	Unstructured	10, 11, 18, 27, 28	17%

The queries were specified to cover the areas of big data analytics as well as the technical dimensions of big data processing. Below is an overview of the business functions as proposed by McKinsey [7] and the associated BigBench queries:

Marketing 18.6%
    Cross-selling 1,2,3,29,30
    Customer micro-segmentation 4,5,6,9,25,26
    Sentiment analysis 8,10,11,18,28
    Enhancing multichannel consumer experiences 12,13
Merchandising 16.7%
    Assortment optimization 14,21,27
    Pricing optimization 16,17
Operations 13.3%
    Performance transparency 7,15
    Customer return analysis 19,20
Supply chain 6.7%
    Inventory management 22,23
New business models 3.3%
    Price comparison 24

In Appendix A, we list all 30 BigBench queries. It has to be noted that some of the queries are identical to TPC-DS queries, this is true for the SQL code as well as the English description. For those queries, we list the original template number in brackets in the description below.

5   Evaluation

We chose to initially run BigBench on the Teradata Aster DBMS. TAD has all features needed to store and process big data. Data can be stored as tables and queries can be executed using the SQL-MR interface that extends declarative SQL with MR processing.

TAD is based on the nCluster technology. nCluster is a shared-nothing parallel database, optimized for data warehousing and analytic workloads [2]. nCluster manages a cluster of commodity server nodes, and is designed to scale out to hundreds of nodes and scale up to petabytes of active data.

The test was executed on a 8 node Teradata Aster appliance. Each node is a Dell server with two quad-core Xeon 5500 at 3.07Ghz and hardware RAID 1 with 8 2.5" drives.

For the test, DSdgen is used to produce the selected TPC-DS tables included in our data model. We used PDGF to generate the additional parts of the data model. The new parts produced by PDGF include the Item_marketprice table, an Apache-style web server log, and the XML configuration for the online review generator. PDGF is also configured to generate references (PK-FK relationships) in the new data that matches the TPC-DS data. In the future, we plan on extending PDGF to handle the generation of TPC-DS tables without the need for DSdgen.

The data was loaded into TAD as tables. The web logs were parsed and converted to a table similar to the structure shown in Appendix B. Product reviews are also interpreted as a table assuming the review text as a VARCHAR(5000).

As a proof of concept, we executed the workload as a single stream without velocity on a ca. 130 GB data set. This corresponds to a scale factor 100 in TPC-DS. Since we adapt the velocity methodology from TPC-DS, it and can easily be implemented with a simple driver that periodically adds data to the system and re-submits a new stream of queries. Furthermore, the addition of concurrent query streams can be handled similarly to benchmarks such as TPC-H.

The query processing times for the individual queries can be seen below.

Query run-time (sec)		Query run-time (sec)	
A1	200	A16	8700.045
A2	12.529	A17	146.879
A3	19.948	A18	1507.33
A4	33.345	A19	11.368
A5	9.462	A20	345
A6	11.652	A21	109.817
A7	1.176	A22	114.555
A8	12.581	A23	1113.373
A9	8.698	A24	11.714
A10	24.847	A25	254.474
A11	2713.042	A26	2708.261
A12	918.575	A27	4.617
A13	1572	A28	381.005
A14	7.952	A29	7.201
A15	41.747	A30	6208

6   Conclusion

In summary, we present the first end-to-end benchmark for big data analytics. While previous work focused on one type of data or processing, we produced 30 queries that address all the three technical dimensions described above. The queries cover all the six major business areas of DB analytics mentioned earlier. We developed and implemented a novel technique for producing unstructured text data and integrated it with traditional structured data generators. We conducted a proof of concept of the proposal by executing it on the Teradata Aster DBMS.

Currently, all queries are translated to the Hadoop eco-system. The complete data generator will be migrated to PDGF, which will make it possible to generate more complex dependencies consistently across the different parts of the schema. This will add correlations that are desirable for exercising analytical queries. Although, basic metrics were specified in [1], we will extend this part of the specification with additional approaches, directly targeting big data related questions.

References

1. Ghazal, A., Rabl, T., Hu, M., Raab, F., Poess, M., Crolotte, A., Jacobsen., H.A.: BigBench: Towards an industry standard benchmark for big data analytics. In: Proceedings of the ACM SIGMOD Conference (2013)
2. Friedman, E., Pawlowski, P., Cieslewicz, J.: SQL/MapReduce: A Practical Approach to Self-Describing, Polymorphic, and Parallelizable User-Defined Functions. PVLDB 2(2), 1402–1413 (2009)
3. Teradata Aster: Teradata Aster Big Analytics Appliance 3H - Analytics Foundation User Guide. Release 5.0.1 edn (2012), http://www.info.teradata.com/edownload.cfm?itemid=123060004
4. Laney, D.: 3D Data Management: Controlling Data Volume, Velocity and Variety. Technical report, Meta Group (2001)
5. Nambiar, R.O., Poess, M.: The Making of TPC-DS. In: VLDB, pp. 1049–1058 (2006)
6. Rabl, T., Frank, M., Sergieh, H.M., Kosch, H.: A Data Generator for Cloud-Scale Benchmarking. In: Nambiar, R., Poess, M. (eds.) TPCTC 2010. LNCS, vol. 6417, pp. 41–56. Springer, Heidelberg (2011)
7. Manyika, J., Chui, M., Brown, B., Bughin, J., Dobbs, R., Roxburgh, C., Byers, A.H.: Big data: The Next Frontier for Innovation, Competition, and Productivity. Technical report, McKinsey Global Institute (2011), http://www.mckinsey.com/insights/mgi/research/technology_and_innovation/big_data_the_next_frontier_for_innovation

A   BigBench Queries

Below all 30 queries of the BigBench proposal are shown. The queries are specified in English, to give a high-level understanding what the business question of each query is. Additionally, an SQL-MR syntax-based description is given [2,3].

Query 1. Find products are sold together frequently in given stores. Only products in certain categories sold in specific stores are considered, and "sold together frequently" means at least 50 customers bought these products together in a transaction.

```
SELECT pid1 AS item1, pid2 AS item2, COUNT(*) AS cnt
  FROM basket_generator (ON
          (SELECT s.ss_ticket_number AS oid, s.ss_item_sk AS pid
             FROM store_sales100 s
             INNER JOIN item100 i ON s.ss_item_sk = i_item_sk
             WHERE i.i_category_id in (1,4,6) and s.ss_store_sk = 10
          PARTITION BY oid
          basket_size(2)
          basket_item('pid')
          item_set_max(500)
          )
  GROUP BY 1,2
HAVING COUNT(pid1) > 49
  ORDER BY 1,3,2;
```

Listing 1.1. Query 1

Query 2. Find the top 30 products that are mostly viewed together with a given product in online store. Note that the order of products viewed does not matter.

```
SELECT pid1 AS item1, pid2 AS item2, COUNT(1) AS cnt
  FROM basket_generator (ON
          (SELECT wcs_user_sk AS cid, wcs_item_sk AS pid
             FROM web_clickstreams
             WHERE wcs_item_sk IS NOT NULL
               AND wcs_user_sk IS NOT NULL
          )
          PARTITION BY cid
          basket_size(2)
          basket_item('pid')
          item_set_max(500)
          )
  WHERE pid1 IN (1416,9082,1547)
  GROUP BY 1,2
  ORDER BY 1,3,2
  LIMIT 30;
```

Listing 1.2. Query 2

Query 3. Find the last 5 products that are mostly viewed before a given product was purchased online. Only products in certain categories and viewed within 10 days before the purchase date are considered.

```
SELECT lastviewed_item, purchased_item, COUNT(*)
  FROM nPath (ON web_clickstreams
          PARTITION BY wcs_user_sk
             ORDER BY wcs_click_date_sk, wcs_click_time_sk
               MODE ('NONOVERLAPPING')
             PATTERN ('A+.B')
             SYMBOLS (true AS A, wcs_sales_sk IS NOT NULL AS B)
               RESULT (
                   LAST (wcs_item_sk OF A) AS lastviewed_item,
                   LAST (wcs_click_date_sk OF A) AS lastviewed_date,
                   FIRST (wcs_item_sk OF B) AS purchased_item,
                   FIRST (wcs_click_date_sk OF B) AS purchased_date
               )
          )
  WHERE purchased_item = 16891
    AND purchased_date - lastviewed_date < 11
  GROUP BY 1,2;
```

Listing 1.3. Query 3

Query 4. Shopping cart abandonment analysis: For users who added products in their shopping carts but did not check out in the online store, find the average number of pages they visited during their sessions.

```
DROP VIEW sessions;

CREATE VIEW sessions AS (
  SELECT *
    FROM sessionize (ON
         (SELECT c.wcs_user_sk as uid, c.wcs_item_sk as item,
                 w.wp_type as wptype,
                 d.d_date + t.t_time*INTERVAL '1 second' as tstamp
            FROM web_clickstreams c, web_page w, date_dim d, time_dim t
           WHERE c.wcs_web_page_sk = w.wp_web_page_sk
             AND c.wcs_click_date_sk = d.d_date_sk
             AND c.wcs_click_time_sk = t.t_time_sk
             AND c.wcs_user_sk IS NOT NULL
         ) AS clicks
         PARTITION BY uid
            ORDER BY tstamp
          timecolumn ('tstamp')
              timeout ('300')
         )
   ORDER BY uid, tstamp
);

DROP VIEW cart_abadon;
CREATE VIEW cart_abadon AS (
  SELECT *
    FROM nPath(ON sessions
       PARTITON BY sessionid
          ORDER BY tstamp
            MODE ('NONOVERLAPPING')
         PATTERN ('C*.A.B*$')
         SYMBOLS (wptype = 'dynamic' AS A, true as C, wptype <> 'order' AS B)
          RESULT (FIRST_NOTNULL (sessionid OF C) AS sid,
                  LAST_NOTNULL (tstamp OF B) AS end_s,
                  FIRST_NOTNULL (tstamp OF C) AS start_s
         )
       )
);

SELECT c.sid, COUNT(*) AS s_pages
  FROM cart_abadon c, sessions s
 WHERE s.sessionid = c.sid
 GROUP BY 1;
```

Listing 1.4. Query 4

Query 5. Build a model using logistic regression: based on existing users online activities and demographics, for a visitor to an online store, predict the visitors likelihood to be interested in a given category.

```
DROP VIEW logstic_reg_t;
CREATE VIEW logstic_reg_t AS (
  SELECT c_customer_sk, college_education, male,
         CASE WHEN clicks_in_category > 2 THEN true ELSE false END AS label
    FROM (
       SELECT c_customer_sk,
              CASE WHEN (cd_education_status = 'Advanced Degree'
                    OR cd_education_status = 'College'
                    OR cd_education_status = '4 yr Degree'
                    OR cd_education_status = '2 yr Degree ')
                THEN TRUE ELSE FALSE END AS college_education,
              CASE WHEN cd_gender = 'M' THEN TRUE ELSE FALSE END AS male,
```

```
             SUM (CASE WHEN i_category='Books' THEN 1 ELSE 0 END) AS
                 clicks_in_category
        FROM customer, customer_demographics, item, web_clickstreams
       WHERE wcs_user_sk = c_customer_sk
         AND c_current_cdemo_sk = cd_demo_sk
         AND wcs_item_sk = i_item_sk
       GROUP BY 1,2,3) C
);

DROP TABLE books_interests;
SELECT *
  FROM log_regression (
    ON (SELECT 1)
    PARTITION BY 1
    DATABASE('benchmark')
    USERID('benchmark')
    PASSWORD('benchmark')
    INPUTTABLE('logstic_reg_t')
    OUTPUTTABLE('books_interests')
    COLUMNNAMES('label','college_education','male')
  );
```

Listing 1.5. Query 5

Query 6. (TPC-DS 4) Find customers who spend more money via web than in stores for a given year. Report customers first name, last name, their country of origin and identify if they are preferred customer.

```
BEGIN;
DROP TABLE IF EXISTS q04_year_total_8;

CREATE TEMP TABLE q04_year_total_8 (
  customer_id            VARCHAR(16),
  customer_first_name    CHAR(20),
  customer_last_name     CHAR(30),
  c_preferred_cust_flag  CHAR(1),
  c_birth_country        VARCHAR(20),
  c_login                CHAR(13),
  c_email_address        CHAR(50),
  dyear                  INTEGER,
  year_total             DECIMAL(15,2),
  sale_type              VARCHAR(2)
) DISTRIBUTE BY HASH (customer_id) AS (
    SELECT c_customer_id::VARCHAR AS customer_id,
           c_first_name           AS customer_first_name,
           c_last_name            AS customer_last_name,
           c_preferred_cust_flag,
           c_birth_country,
           c_login,
           c_email_address,
           sv.d_year              AS dyear,
           sv.year_total          AS year_total,
           's'::VARCHAR           AS sale_type
      FROM customer,
        (SELECT ss.ss_customer_sk AS customer_sk,
                dt.d_year          AS d_year,
                SUM(((ss_ext_list_price - ss_ext_wholesale_cost
                    - ss_ext_discount_amt) + ss_ext_sales_price) / 2)
                  AS year_total
           FROM store_sales ss, date_dim dt
          WHERE ss.ss_sold_date_sk = dt.d_date_sk
          GROUP BY ss.ss_customer_sk, dt.d_year) sv
    WHERE c_customer_sk = sv.customer_sk
    UNION ALL
    SELECT c_customer_id::VARCHAR AS customer_id,
           c_first_name           AS customer_first_name,
```

```
                c_last_name                AS customer_last_name,
                c_preferred_cust_flag,
                c_birth_country,
                c_login,
                c_email_address,
                cv.d_year                  AS dyear,
                cv.year_total              AS year_total,
                'c'::VARCHAR               AS sale_type
          FROM customer,
                (SELECT ws.ws_bill_customer_sk AS customer_sk
                        dt.d_year               AS d_year
                        SUM(((ws_ext_list_price - ws_ext_wholesale_cost
                              - ws_ext_discount_amt) + ws_ext_sales_price) / 2)
                                        AS year_total
                   FROM web_sales ws,
                        date_dim dt
                  WHERE ws.ws_sold_date_sk = dt.d_date_sk
                  GROUP BY ws.ws_bill_customer_sk, dt.d_year) cv
          WHERE c_customer_sk = cv.customer_sk);

ANALYZE q04_year_total_8;

SELECT t_s_secyear.customer_id,
       t_s_secyear.customer_first_name,
       t_s_secyear.customer_last_name,
       t_s_secyear.c_preferred_cust_flag,
       t_s_secyear.c_birth_country,
       t_s_secyear.c_login
  FROM q04_year_total_8 t_s_firstyear,
       q04_year_total_8 t_s_secyear,
       q04_year_total_8 t_c_firstyear,
       q04_year_total_8 t_c_secyear
 WHERE t_s_secyear.customer_id = t_s_firstyear.customer_id
   AND t_s_firstyear.customer_id = t_c_secyear.customer_id
   AND t_s_firstyear.customer_id = t_c_firstyear.customer_id
   AND t_s_firstyear.sale_type   = 's'
   AND t_c_firstyear.sale_type   = 'c'
   AND t_s_secyear.sale_type     = 's'
   AND t_c_secyear.sale_type     = 'c'
   AND t_s_firstyear.dyear       = 1999
   AND t_s_secyear.dyear         = 1999 + 1
   AND t_c_firstyear.dyear       = 1999
   AND t_c_secyear.dyear         = 1999 + 1
   AND t_s_firstyear.year_total  > 0
   AND t_c_firstyear.year_total  > 0
   AND CASE WHEN t_c_firstyear.year_total > 0
            THEN t_c_secyear.year_total / t_c_firstyear.year_total
            ELSE NULL END >
       CASE WHEN t_s_firstyear.year_total > 0
            THEN t_s_secyear.year_total / t_s_firstyear.year_total
            ELSE NULL END
 ORDER BY t_s_secyear.customer_id,
          t_s_secyear.customer_first_name,
          t_s_secyear.customer_last_name,
          t_s_secyear.c_preferred_cust_flag,
          t_s_secyear.c_birth_country,
          t_s_secyear.c_login
 LIMIT 100;

DROP TABLE IF EXISTS q04_year_total_8;
END;
```

Listing 1.6. Query 6

Query 7. (TPC-DS 6) List all the stores with at least 10 customers who during a given month bought products with the price tag at least 20% higher than the average price of products in the same category.

```
BEGIN;
DROP TABLE IF EXISTS q06_specific_month_88;
DROP TABLE IF EXISTS q06_cat_avg_price_88;

CREATE DIMENSION TABLE q06_specific_month_88 AS
  SELECT DISTINCT (d_month_seq) AS d_month_seq
    FROM date_dim
    WHERE d_year = 2002
      AND d_moy = 7;

CREATE DIMENSION TABLE q06_cat_avg_price_88 AS
    SELECT i_category AS i_category,
        AVG (i_current_price) * 1.2 AS avg_price
      FROM item
      GROUP BY i_category;

SELECT a.ca_state AS state, count(*) as cnt
  FROM customer_address a, customer c,
        store_sales s, date_dim d, item i,
            q06_specific_month_88 m, q06_cat_avg_price_88  p
  WHERE a.ca_address_sk   = c.c_current_addr_sk
    AND c.c_customer_sk   = s.ss_customer_sk
    AND s.ss_sold_date_sk = d.d_date_sk
    AND s.ss_item_sk      = i.i_item_sk
    AND d.d_month_seq     = m.d_month_seq
    AND p.i_category      = i.i_category
    AND i.i_current_price > p.avg_price
  GROUP BY a.ca_state
  HAVING COUNT(*) >= 10
  ORDER BY cnt
  LIMIT 100;

DROP TABLE IF EXISTS q06_specific_month_88;
DROP TABLE IF EXISTS q06_cat_avg_price_88;
END;
```

Listing 1.7. Query 7

Query 8. For online sales, compare the total sales in which customers checked online reviews before making the purchase and that of sales in which customers did not read reviews. Consider only online sales for a specific category in a given year.

```
BEGIN;
DROP VIEW clicks;
CREATE VIEW clicks AS (
  SELECT c.wcs_item_sk AS item,
         c.wcs_user_sk AS uid,
         c.wcs_click_date_sk AS c_date,
         c.wcs_click_time_sk AS c_time,
         c.wcs_sales_sk AS sales_sk,
         w.wp_type AS wpt
    FROM web_clickstreams c, web_page w
   WHERE c.wcs_web_page_sk = w.wp_web_page_sk
     and c.wcs_user_sk IS NOT NULL
);

DROP VIEW sales_review;
CREATE VIEW sales_review AS (
  SELECT s_sk
```

```
    FROM nPath(ON clicks
            PARTITION BY uid
            ORDER BY c_date, c_time
            MODE ('NONOVERLAPPING')
            PATTERN ('A+.C*.B')
            SYMBOLS (wpt = 'review' AS A, TRUE AS C,
                    sales_sk IS NOT NULL AS B)
            RESULT (FIRST (c_date OF B) AS s_date,
                    FIRST (sales_sk OF B) AS s_sk))
    WHERE s_date > 2451424 AND s_date <2451424+365
);

SELECT SUM (CASE WHEN ws.ws_sk IN (SELECT * FROM sales_review)
                THEN ws_net_paid
                ELSE 0 END) AS review_sales_amount,
        SUM (ws_net_paid) -
        SUM (CASE WHEN ws.ws_sk IN (SELECT * FROM sales_review)
                THEN ws_net_paid
                ELSE 0 END) AS no_review_sales_amount
    FROM web_sales ws
WHERE ws.ws_sold_date_sk > 2451424
    AND ws.ws_sold_date_sk <2451424+365;
END;
```

Listing 1.8. Query 8

Query 9. (TPC-DS 48) Calculate the total sales by different types of customers (e.g., based on marital status, education status), sales price and different combinations of state and sales profit.

```
SELECT SUM (ss_quantity)
    FROM store_sales, store, customer_demographics,
        customer_address, date_dim
    WHERE s_store_sk = ss_store_sk
    AND   ss_sold_date_sk = d_date_sk
    AND d_year = 1998
    AND ((cd_demo_sk = ss_cdemo_sk
        AND cd_marital_status = 'M'
        AND cd_education_status = '4 yr Degree'
        AND ss_sales_price between 100.00 AND 150.00)
        OR
          (cd_demo_sk = ss_cdemo_sk
        AND cd_marital_status = 'M'
        AND cd_education_status = '4 yr Degree'
        AND ss_sales_price between 50.00 AND 100.00)
        OR
          (cd_demo_sk = ss_cdemo_sk
        AND cd_marital_status = 'M'
        AND cd_education_status = '4 yr Degree'
        AND ss_sales_price between 150.00 AND 200.00))
    AND ((ss_addr_sk = ca_address_sk
        AND ca_country = 'United States'
        AND ca_state in ('KY', 'GA', 'NM')
        AND ss_net_profit between 0 AND 2000)
        OR
          (ss_addr_sk = ca_address_sk
        AND ca_country = 'United States'
        AND ca_state in ('MT', 'OR', 'IN')
        AND ss_net_profit between 150 AND 3000)
        OR
          (ss_addr_sk = ca_address_sk
        AND ca_country = 'United States'
        AND ca_state in ('WI', 'MO', 'WV')
        AND ss_net_profit between 50 AND 25000));
```

Listing 1.9. Query 9

Query 10. For all products, extract sentences from its product reviews that contain positive or negative sentiment and display the sentiment polarity of the extracted sentences.

```
SELECT pr_item_sk, out_content, out_polarity, out_sentiment_words
  FROM ExtractSentiment
       (ON product_reviews100
        TEXT_COLUMN ('pr_review_content')
        MODEL ('dictionary')
        LEVEL ('sentence')
        ACCUMULATE ('pr_item_sk')
       )
 WHERE out_polarity = 'NEG'
    OR out_polarity = 'POS';
```
Listing 1.10. Query 10

Query 11. For a given product, measure the correlation of sentiments, including the number of reviews and average review ratings, on product monthly revenues.

```
BEGIN;
DROP VIEW IF EXISTS review_stats;
CREATE VIEW review_stats AS(
  SELECT p.pr_item_sk AS pid,
         CAST(p.r_count AS INT) AS reviews_count,
         CAST(p.avg_rating AS INT) AS avg_rating,
         CAST(s.revenue AS INT) AS m_revenue
    FROM (SELECT pr_item_sk, COUNT(*) AS r_count,
                 AVG(pr_review_rating) AS avg_rating
            FROM product_reviews
           WHERE pr_item_sk IS NOT NULL
           GROUP BY 1) p
         JOIN
         (SELECT ws_item_sk, SUM(ws_net_paid) AS revenue
            FROM web_sales
           WHERE ws_sold_date_sk > 2452642-30
             AND ws_sold_date_sk < 2452642
             AND ws_item_sk IS NOT NULL
           GROUP BY 1) s
         ON p.pr_item_sk = s.ws_item_sk);

SELECT *
  FROM corr_reduce (ON
         corr_map (ON
           review_stats
           COLUMNS ('[m_revenue:reviews_count],[m_revenue:avg_rating]')
           KEY_NAME('k')
         )
         PARTITION BY k);

DROP VIEW review_stats;
END;
```
Listing 1.11. Query 11

Query 12. Find all customers, who viewed items of a given category on the web in a given month and year that was followed by an in-store purchase in the three consecutive months.

```
SELECT *
  FROM nPath (
    ON (SELECT c.wcs_item_sk AS item,
               c.wcs_user_sk AS uid,
```

```
              c.wcs_click_date_sk AS c_date,
              c.wcs_click_time_sk AS c_time
        FROM web_clickstreams c, item i
       WHERE c.wcs_item_sk = i.i_item_sk
         AND i.i_category in ('Books', 'Electronics')
         AND c.wcs_user_sk IS NOT NULL
         AND c.wcs_click_date_sk > 2451424
         AND c.wcs_click_date_sk < 2451424+30) AS click
    PARTITION BY uid
    ORDER BY c_date, c_time
    ON (SELECT s.ss_item_sk AS item,
              s.ss_customer_sk AS uid,
              s.ss_sold_date_sk AS s_date,
              s.ss_sold_time_sk AS s_time
        FROM store_sales s, item i
       WHERE s.ss_item_sk = i.i_item_sk
         AND i.i_category in ('Books', 'Electronics')
         AND s.ss_customer_sk IS NOT NULL
         AND s.ss_sold_date_sk > 2451424
         AND s.ss_sold_time_sk < 2451424+120) AS sale
    PARTITION BY uid order by s_date, s_time
    MODE ('NONOVERLAPPING')
    PATTERN ('(c+).(s)')
    SYMBOLS (click.uid IS NOT NULL AS c,
            sale.uid IS NOT NULL AS s)
    RESULT (FIRST(c_date OF c) AS c_date,
            FIRST(s_date OF s) AS s_date,
            FIRST(sale.uid OF s) AS user_sk)
);
```

Listing 1.12. Query 12

Query 13. (TPC-DS 74) Display customers with both store and web sales in consecutive years for whom the increase in web sales exceeds the increase in store sales for a specified year.

```
BEGIN;
DROP TABLE IF EXISTS q74_customer_year_total_880;

CREATE TEMP TABLE q74_customer_year_total_880(
  customer_id           VARCHAR(16),
  customer_first_name   CHAR(20)
  customer_last_name    CHAR(30)
  year                  INTEGER
  year_total            DECIMAL(15,2)
  sale_type             VARCHAR(2))
  DISTRIBUTE BY hash (customer_id) AS
  SELECT c_customer_id     customer_id,
         c_first_name      customer_first_name,
         c_last_name       customer_last_name,
         d_year            year,
         SUM(ss_net_paid)  year_total,
         's'::VARCHAR      sale_type
    FROM customer, store_sales, date_dim
   WHERE c_customer_sk = ss_customer_sk
     AND ss_sold_date_sk = d_date_sk
     AND d_year IN (1999 ,1999 + 1)
   GROUP BY c_customer_id, c_first_name,
            c_last_name, d_year
  UNION ALL
  SELECT c_customer_id     customer_id,
         c_first_name      customer_first_name,
         c_last_name       customer_last_name,
         d_year            year,
         SUM(ws_net_paid)  year_total,
         'w'::VARCHAR      sale_type
```

```
       FROM customer, web_sales, date_dim
      WHERE c_customer_sk = ws_bill_customer_sk
        AND ws_sold_date_sk = d_date_sk
        AND d_year IN (1999 ,1999 + 1)
      GROUP BY c_customer_id, c_first_name,
               c_last_name, d_year;

   SELECT t_s_secyear.customer_id, t_s_secyear.customer_first_name,
          t_s_secyear.customer_last_name
     FROM q74_customer_year_total_880 t_s_firstyear,
          q74_customer_year_total_880 t_s_secyear,
          q74_customer_year_total_880 t_w_firstyear,
          q74_customer_year_total_880 t_w_secyear
    WHERE t_s_secyear.customer_id   = t_s_firstyear.customer_id
      AND t_s_firstyear.customer_id = t_w_secyear.customer_id
      AND t_s_firstyear.customer_id = t_w_firstyear.customer_id
      AND t_s_firstyear.sale_type   = 's'
      AND t_w_firstyear.sale_type   = 'w'
      AND t_s_secyear.sale_type     = 's'
      AND t_w_secyear.sale_type     = 'w'
      AND t_s_firstyear.year        = 1999
      AND t_s_secyear.year          = 1999 + 1
      AND t_w_firstyear.year        = 1999
      AND t_w_secyear.year          = 1999 + 1
      AND t_s_firstyear.year_total  > 0
      AND t_w_firstyear.year_total  > 0
      AND CASE WHEN t_w_firstyear.year_total > 0
               THEN t_w_secyear.year_total / t_w_firstyear.year_total
               ELSE NULL END
        > CASE WHEN t_s_firstyear.year_total > 0
               THEN t_s_secyear.year_total / t_s_firstyear.year_total
               ELSE NULL END
   ORDER BY 1
   LIMIT 100;

   DROP TABLE IF EXISTS q74_customer_year_total_880;
   END;
```

Listing 1.13. Query 13

Query 14. (TPC-DS 90) What is the ratio between the number of items sold over the internet in the morning (8 to 9am) to the number of items sold in the evening (7 to 8pm) of customers with a specified number of dependents. Consider only websites with a high amount of content.

```
SELECT CAST(amc AS DECIMAL(15,4)) / CAST(pmc AS DECIMAL(15,4)) am_pm_ratio
   FROM (SELECT COUNT(*) amc
           FROM web_sales, household_demographics, time_dim, web_page wp
          WHERE ws_sold_time_sk = time_dim.t_time_sk
            AND ws_ship_hdemo_sk = household_demographics.hd_demo_sk
            AND ws_web_page_sk = wp.wp_web_page_sk
            AND time_dim.t_hour BETWEEN 8 AND 8+1
            AND household_demographics.hd_dep_count = 5
            AND wp.wp_char_count BETWEEN 5000 AND 5200) at,
        (SELECT COUNT(*) pmc
           FROM web_sales, household_demographics , time_dim, web_page wp
          WHERE ws_sold_time_sk = time_dim.t_time_sk
            AND ws_ship_hdemo_sk = household_demographics.hd_demo_sk
            AND ws_web_page_sk = wp.wp_web_page_sk
            AND time_dim.t_hour BETWEEN 19 AND 19+1
            AND household_demographics.hd_dep_count = 5
            AND wp.wp_char_count BETWEEN 5000 AND 5200) pt
   ORDER BY am_pm_ratio ;
```

Listing 1.14. Query 14

Query 15. Find the categories with flat or declining sales for in store purchases during a given year for a given store.

```
BEGIN;
DROP VIEW IF EXISTS category_coefficient;
DROP VIEW IF EXISTS time_series_category;

CREATE VIEW time_series_category AS (
  SELECT i.i_category_id AS cat,
         s.ss_sold_date_sk AS d,
         SUM(s.ss_net_paid) AS sales
    FROM store_sales s, item i
   WHERE s.ss_item_sk = i.i_item_sk
     AND i.i_category_id IS NOT NULL
     AND s.ss_sold_date_sk > 2451424
     AND s.ss_sold_date_sk < 2451424+365
     AND s.ss_store_sk = 10
   GROUP BY 1,2
);

CREATE VIEW category_coefficient AS (
  SELECT 1 AS category, coefficient_index, value AS slope
    FROM linreg (ON
           linregmatrix (ON
             (SELECT d, sales
                FROM time_series_category
               WHERE cat = 1)
           ) PARTITION BY 1
         )
   WHERE coefficient_index = 1
  UNION ALL
  SELECT 2, coefficient_index, value
    FROM linreg (ON
           linregmatrix (ON
             (SELECT d, sales
                FROM time_series_category
               WHERE cat = 2)
           ) PARTITION BY 1
         )
   WHERE coefficient_index = 1
  UNION ALL
  SELECT 3, coefficient_index, value
    FROM linreg (ON
           linregmatrix (ON
             (SELECT d, sales
                FROM time_series_category
               WHERE cat = 3)
           ) PARTITION BY 1
         )
   WHERE coefficient_index = 1
  UNION ALL
  SELECT 4, coefficient_index, value
    FROM linreg (ON
           linregmatrix (ON
             (SELECT d, sales
                FROM time_series_category
               WHERE cat = 4)
           ) PARTITION BY 1
         )
   WHERE coefficient_index = 1
  UNION ALL
  SELECT 5, coefficient_index, value
    FROM linreg (ON
           linregmatrix (ON
             (SELECT d, sales
                FROM time_series_category
               WHERE cat = 5)
           ) PARTITION BY 1
```

```
            )
  WHERE coefficient_index = 1
UNION ALL
SELECT 6, coefficient_index, value
   FROM linreg (ON
            linregmatrix (ON
              (SELECT d, sales
                  FROM time_series_category
                  WHERE cat = 6)
            ) PARTITION BY 1
         )
  WHERE coefficient_index = 1
UNION ALL
SELECT 7, coefficient_index, value
   FROM linreg (ON
            linregmatrix (ON
              (SELECT d, sales
                  FROM time_series_category
                  WHERE cat = 7)
            ) PARTITION BY 1
         )
  WHERE coefficient_index = 1
UNION ALL
SELECT 8, coefficient_index, value
   FROM linreg (ON
            linregmatrix (ON
              (SELECT d, sales
                  FROM time_series_category
                  WHERE cat = 8)
            ) partition by 1
         )
  WHERE coefficient_index = 1
UNION ALL
SELECT 9, coefficient_index, value
   FROM linreg (ON
            linregmatrix (ON
              (SELECT d, sales
                  FROM time_series_category
                  WHERE cat = 9)
            ) partition by 1
         )
  WHERE coefficient_index = 1
UNION ALL
SELECT 10, coefficient_index, value
   FROM linreg (ON
            linregmatrix (ON
              (SELECT d, sales
                  FROM time_series_category
                  WHERE cat = 10)
            ) partition by 1
         )
  WHERE coefficient_index = 1;

SELECT * FROM category_coefficient WHERE slope < 0;

DROP VIEW category_coefficient;
DROP VIEW time_series_category;
END;
```

Listing 1.15. Query 15

Query 16. (TPC-DS 40) Compute the impact of an item price change on the store sales by computing the total sales for items in a 30 day period before and after the price change. Group the items by location of warehouse where they were delivered from.

```
SELECT w_state, i_item_id,
       SUM (CASE WHEN (CAST (d_date AS DATE) < CAST ('1998-03-16' AS DATE))
                 THEN ws_sales_price - coalesce(wr_refunded_cash,0) ELSE 0
                   END)
                 AS sales_before,
       SUM (CASE WHEN (CAST (d_date AS DATE) >= CAST ('1998-03-16' AS DATE))
                 THEN ws_sales_price - coalesce(wr_refunded_cash,0) ELSE 0
                   END)
                 AS sales_after
   FROM web_sales LEFT OUTER JOIN web_returns
        ON (ws_order_number = wr_order_number
        AND ws_item_sk = wr_item_sk),
        warehouse, item, date_dim
  WHERE i_item_sk        = ws_item_sk
    AND ws_warehouse_sk = w_warehouse_sk
    AND ws_sold_date_sk = d_date_sk
    AND d_date BETWEEN (CAST ('1998-03-16' AS DATE) - INTERVAL '30 day')
                   AND (CAST ('1998-03-16' AS DATE) + INTERVAL '30 day')
  GROUP BY w_state,i_item_id
  ORDER BY w_state,i_item_id;
```

Listing 1.16. Query 16

Query 17. (TPC-DS 61) Find the ratio of items sold with and without promotions in a given month and year. Only items in certain categories sold to customers living in a specific time zone are considered.

```
SELECT promotions, total,
       CAST(promotions AS DECIMAL(15,4)) /
       CAST(total AS DECIMAL(15,4)) * 100
  FROM (SELECT SUM (ss_ext_sales_price) promotions
          FROM store_sales, store, promotion, date_dim,
               customer, customer_address, item
         WHERE ss_sold_date_sk = d_date_sk
           AND ss_store_sk = s_store_sk
           AND ss_promo_sk = p_promo_sk
           AND ss_customer_sk= c_customer_sk
           AND ca_address_sk = c_current_addr_sk
           AND ss_item_sk = i_item_sk
           AND ca_gmt_offset = -7
           AND i_category = 'Jewelry'
           AND (p_channel_dmail = 'Y' OR p_channel_email = 'Y'
                                      OR p_channel_tv = 'Y')
           AND s_gmt_offset = -7
           AND d_year = 2001
           AND d_moy   = 12) promotional_sales,
        (SELECT sum(ss_ext_sales_price) total
          FROM store_sales, store, date_dim,
               customer, customer_address, item
         WHERE ss_sold_date_sk = d_date_sk
           AND ss_store_sk = s_store_sk
           AND ss_customer_sk= c_customer_sk
           AND ca_address_sk = c_current_addr_sk
           AND ss_item_sk = i_item_sk
           AND ca_gmt_offset = -7
           AND i_category = 'Jewelry'
           AND s_gmt_offset = -7
           AND d_year = 2001
           AND d_moy   = 12) all_sales
  ORDER BY promotions, total;
```

Listing 1.17. Query 17

Query 18. Identify the stores with flat or declining sales in 3 consecutive months, check if there are any negative reviews regarding these stores available online.

```
BEGIN;
DROP VIEW IF EXISTS store_coefficient;
DROP VIEW IF EXISTS time_series_store;

CREATE VIEW time_series_store AS (
  SELECT ss_store_sk AS store, ss_sold_date_sk AS d,
         SUM(ss_net_paid) AS sales
    FROM store_sales
   WHERE ss_sold_date_sk > 2451424
     AND ss_sold_date_sk < 2451424+90
   GROUP BY 1,2);

CREATE VIEW store_coefficient AS (
  SELECT 1 AS store, coefficient_index, value AS slope
    FROM linreg (ON
            linregmatrix (ON
              (SELECT d, sales
                 FROM time_series_store
                WHERE store = 1)
            ) PARTITION BY 1
         )
  WHERE coefficient_index = 1
UNION ALL
SELECT 2 AS store, coefficient_index, value AS slope
   FROM linreg (ON
            linregmatrix (ON
              (SELECT d, sales
                 FROM time_series_store
                WHERE store = 2)
            ) PARTITION BY 1
         )
  WHERE coefficient_index = 1
UNION ALL
SELECT 3 AS store, coefficient_index, value AS slope
   FROM linreg (ON
            linregmatrix (ON
              (SELECT d, sales
                 FROM time_series_store
                WHERE store = 3)
            ) PARTITION BY 1
         )
  WHERE coefficient_index = 1
UNION ALL
SELECT 4 AS store, coefficient_index, value AS slope
   FROM linreg (ON
            linregmatrix (ON
              (SELECT d, sales
                 FROM time_series_store
                WHERE store = 4)
            ) PARTITION BY 1
         )
  WHERE coefficient_index = 1
UNION ALL
SELECT 5 AS store, coefficient_index, value AS slope
   FROM linreg (ON
            linregmatrix (ON
              (SELECT d, sales
                 FROM time_series_store
                WHERE store = 5)
            ) PARTITION BY 1
         )
  WHERE coefficient_index = 1
UNION ALL
SELECT 6 AS store, coefficient_index, value AS slope
   FROM linreg (ON
```

```
            linregmatrix (ON
               (SELECT d, sales
                   FROM time_series_store
                   WHERE store = 6)
               ) PARTITION BY 1
          )
   WHERE coefficient_index = 1
UNION ALL
SELECT 7 AS store, coefficient_index, value AS slope
   FROM linreg (ON
            linregmatrix (ON
               (SELECT d, sales
                   FROM time_series_store
                   WHERE store = 7)
               ) PARTITION BY 1
          )
   WHERE coefficient_index = 1
UNION ALL
SELECT 8 AS store, coefficient_index, value AS slope
   FROM linreg (ON
            linregmatrix (ON
               (SELECT d, sales
                   FROM time_series_store
                   WHERE store = 8)
               ) PARTITION BY 1
          )
   WHERE coefficient_index = 1
UNION ALL
SELECT 9 AS store, coefficient_index, value AS slope
   FROM linreg (ON
            linregmatrix (ON
               (SELECT d, sales
                   FROM time_series_store
                   WHERE store = 9)
               ) PARTITION BY 1
          )
   WHERE coefficient_index = 1
UNION ALL
SELECT 10 AS store, coefficient_index, value AS slope
   FROM linreg (ON
            linregmatrix (ON
               (SELECT d, sales
                   FROM time_series_store
                   WHERE store = 10)
               ) PARTITION BY 1
          )
   WHERE coefficient_index = 1
UNION ALL
SELECT 11 AS store, coefficient_index, value AS slope
   FROM linreg (ON
            linregmatrix (ON
               (SELECT d, sales
                   FROM time_series_store
                   WHERE store = 11)
               ) PARTITION BY 1
          )
   WHERE coefficient_index = 1
UNION ALL
SELECT 12 AS store, coefficient_index, value AS slope
   FROM linreg (ON
            linregmatrix (ON
               (SELECT d, sales
                   FROM time_series_store
                   WHERE store = 12)
               ) PARTITION BY 1
          )
   WHERE coefficient_index = 1);
```

```
SELECT s_store_name, pr_review_date, out_content,
       out_polarity, out_sentiment_words
  FROM ExtractSentiment (ON
          (SELECT s_store_name, pr_review_content, pr_review_date
             FROM store_coefficient c, store s, product_reviews
            WHERE c.slope < 0
              AND s.s_store_sk = c.store
              AND pr_review_content like '%'||s_store_name||'%')
          TEXT_COLUMN('pr_review_content')
          MODEL('dictionary')
          LEVEL('DOCUMENT')
          ACCUMULATE('s_store_name','pr_review_date'))
 WHERE out_polarity = 'NEG';

DROP VIEW store_coefficient;
DROP VIEW time_series_store;
END;
```

Listing 1.18. Query 18

Query 19. Retrieve the items with the highest number of returns where the number of returns was approximately equivalent across all store and web channels (within a tolerance of +/- 10%), within the week ending a given date. Analyze the online reviews for these items to see if there are any major negative reviews.

```
BEGIN;

CREATE VIEW sr_items AS
  (SELECT i_item_sk item_id,
          SUM(sr_return_quantity) sr_item_qty
     FROM store_returns, item, date_dim
    WHERE sr_item_sk = i_item_sk
      AND d_date IN
          (SELECT d_date
             FROM date_dim
            WHERE d_week_seq IN
                  (SELECT d_week_seq
                     FROM date_dim
                    WHERE d_date IN
                          ('1998-01-02','1998-10-15','1998-11-10')))
      AND sr_returned_date_sk = d_date_sk
    GROUP BY i_item_sk
   HAVING SUM (sr_return_quantity) > 0);

CREATE VIEW wr_items AS
  (SELECT i_item_sk item_id, SUM(wr_return_quantity) wr_item_qty
     FROM web_returns, item, date_dim
    WHERE wr_item_sk = i_item_sk
      AND d_date IN (SELECT d_date
    FROM date_dim
   WHERE d_week_seq in
         (SELECT d_week_seq
            FROM date_dim
           WHERE d_date IN ('2001-03-10' ,'2001-08-04' ,'2001-11-14')))
      AND wr_returned_date_sk  = d_date_sk
    GROUP BY i_item_sk
   HAVING SUM(wr_return_quantity) > 0);

CREATE VIEW return_items AS
  (SELECT sr_items.item_id item, sr_item_qty,
          100.0 * sr_item_qty / (sr_item_qty + wr_item_qty) / 2.0 sr_dev,
          wr_item_qty, 100.0 * wr_item_qty /
          (sr_item_qty + wr_item_qty) / 2.0 wr_dev,
          (sr_item_qty  +wr_item_qty) / 2.0 "average"
     FROM sr_items, wr_items
```

```
      WHERE sr_items.item_id = wr_items.item_id
      ORDER BY average DESC
      LIMIT 100) ;

SELECT pr_item_sk, out_content, out_polarity, out_sentiment_words
  FROM ExtractSentiment (ON
           product_reviews
           TEXT_COLUMN ('pr_review_content')
           MODEL ('dictionary')
           LEVEL ('sentence')
           ACCUMULATE ('pr_item_sk')
         )
  WHERE out_polarity = 'NEG'
    AND pr_item_sk IN (SELECT item FROM return_items);

DROP VIEW return_items;
DROP VIEW wr_items;
DROP VIEW sr_items;
END;
```

Listing 1.19. Query 19

Query 20. Customer segmentation for return analysis: Customers are separated along the following dimensions: return frequency, return order ratio (total number of orders partially or fully returned versus the total number of orders), return item ratio (total number of items returned versus the number of items purchased), return amount ration (total monetary amount of items returned versus the amount purchased), return order ratio. Consider the store returns during a given year for the computation.

```
CREATE VIEW sales_returns AS (
  SELECT s.ss_sold_date_sk AS s_date,
         r.sr_returned_date_sk AS r_date,
         s.ss_item_sk AS item,
         s.ss_ticket_number AS oid,
         s.ss_net_paid AS s_amount,
         r.sr_return_amt AS r_amount,
         (CASE WHEN s.ss_customer_sk IS NULL
               THEN r.sr_customer_sk ELSE s.ss_customer_sk END) AS cid,
         s.ss_customer_sk AS s_cid,
         sr_customer_sk AS r_cid
    FROM store_sales s LEFT JOIN store_returns100 r ON
             s.ss_item_sk = r.sr_item_sk
         AND s.ss_ticket_number = r.sr_ticket_number
   WHERE s.ss_sold_date_sk IS NOT NULL);

CREATE VIEW clusters AS (
  SELECT cid,
         100.0 * COUNT (DISTINCT (CASE WHEN r_date IS NOT NULL
                                       THEN oid ELSE NULL END))
             / COUNT (DISTINCT oid) AS r_order_ratio,
         SUM (CASE WHEN r_date IS NOT NULL THEN 1 ELSE 0 END)
             / COUNT (item) * 100 AS r_item_ratio,
         SUM (CASE WHEN r_date IS NOT NULL THEN r_amount ELSE 0 END)
             / SUM (s_amount) * 100 AS r_amount_ratio,
         COUNT (DISTINCT (CASE WHEN r_date IS NOT NULL
                               THEN r_date ELSE NULL END))
                               AS r_freq
    FROM sales_returns
   WHERE cid IS NOT NULL
   GROUP BY 1
   HAVING COUNT (DISTINCT (CASE WHEN r_date IS NOT NULL
                               THEN r_date ELSE NULL END)) > 1);
```

```
SELECT *
  FROM kmeans (ON
       (SELECT 1)
       PARTITION BY 1
       DATABASE ('benchmark')
       USERID ('benchmark')
       PASSWORD ('benchmark')
       INPUTTABLE ('clusters␣AS␣c')
       OUTPUTTABLE ('user_return_groups')
       NUMBERK('4'));

SELECT clusterid, cid
  FROM kmeansplot (ON
       clusters AS c
       PARTITION BY ANY
       ON user_return_groups dimension
       CENTROIDSTABLE ('user_return_groups'))
 ORDER BY clusterid, cid;

DROP TABLE user_return_groups;
DROP VIEW clusters;
DROP VIEW sales_returns;
```

Listing 1.20. Query 20

Query 21. (TPC-DS 29) Get all items that were sold in stores in a given month and year and which were returned in the next six months and re-purchased by the returning customer afterwards through the web sales channel in the following three years. For those these items, compute the total quantity sold through the store, the quantity returned and the quantity purchased through the web. Group this information by item and store.

```
SELECT i_item_id, i_item_desc, s_store_id, s_store_name,
       sum(ss_quantity) AS store_sales_quantity,
       sum(sr_return_quantity) AS store_returns_quantity,
       sum(ws_quantity) AS web_sales_quantity
  FROM store_sales, store_returns, web_sales, date_dim d1,
       date_dim d2, date_dim d3, store, item
 WHERE d1.d_moy             = 4
   AND d1.d_year            = 1998
   AND d1.d_date_sk         = ss_sold_date_sk
   AND i_item_sk            = ss_item_sk
   AND s_store_sk           = ss_store_sk
   AND ss_customer_sk       = sr_customer_sk
   AND ss_item_sk           = sr_item_sk
   AND ss_ticket_number     = sr_ticket_number
   AND sr_returned_date_sk  = d2.d_date_sk
   AND d2.d_moy             BETWEEN 4 AND  4 + 3
   AND d2.d_year            = 1998
   AND sr_customer_sk       = ws_bill_customer_sk
   AND sr_item_sk           = ws_item_sk
   AND ws_sold_date_sk      = d3.d_date_sk
   AND d3.d_year            IN (1998,1998+1,1998+2)
 GROUP BY i_item_id, i_item_desc, s_store_id, s_store_name
 ORDER BY i_item_id, i_item_desc, s_store_id, s_store_name;
```

Listing 1.21. Query 21

Query 22. (TPC-DS 21) For all items whose price was changed on a given date, compute the percentage change in inventory between the 30-day period before the price change and the 30-day period after the change. Group this information by warehouse.

```
SELECT *
  FROM (SELECT w_warehouse_name , i_item_id,
               SUM (CASE WHEN (CAST (d_date AS DATE) < CAST ('2000-05-08' AS
                    DATE))
                         THEN inv_quantity_on_hand
                         ELSE 0 END) AS inv_before
               SUM (CASE WHEN (CAST (d_date AS date) >= CAST ('2000-05-08'
                    AS DATE))
                         THEN inv_quantity_on_hand
                         ELSE 0 END) AS inv_after
          FROM inventory, warehouse, item, date_dim
         WHERE i_current_price BETWEEN 0.99 AND 1.49
           AND i_item_sk = inv_item_sk
           AND inv_warehouse_sk = w_warehouse_sk
           AND inv_date_sk = d_date_sk
           AND d_date BETWEEN (CAST ('2000-05-08' AS DATE) - 30)
                          AND (CAST ('2000-05-08' AS DATE) + 30)
         GROUP BY w_warehouse_name , i_item_id) x
 WHERE (CASE WHEN inv_before > 0
             THEN inv_after / inv_before
             ELSE NULL END) BETWEEN 2.0/3.0 AND 3.0/2.0
 ORDER BY w_warehouse_name , i_item_id;
```

Listing 1.22. Query 22

Query 23. (TPC-DS 39) This query contains multiple, related iterations:

1. Calculate the coefficient of variation and mean of every item and warehouse of two consecutive months.
2. Find items that had a coefficient of variation in the first months of 1.5 or larger.

```
BEGIN;

CREATE VIEW inv AS
  (SELECT w_warehouse_name , w_warehouse_sk , i_item_sk,
          d_moy, stdev, mean,
          CASE mean WHEN 0 THEN NULL ELSE stdev/mean END cov
     FROM (SELECT w_warehouse_name , w_warehouse_sk , i_item_sk,
                  d_moy, stddev_samp(inv_quantity_on_hand) stdev,
                  avg(inv_quantity_on_hand) mean
             FROM inventory, item, warehouse, date_dim
            WHERE inv_item_sk = i_item_sk
              AND inv_warehouse_sk = w_warehouse_sk
              AND inv_date_sk = d_date_sk
              AND d_year = 1998
            GROUP BY w_warehouse_name , w_warehouse_sk ,
                     i_item_sk, d_moy) foo
    WHERE CASE mean WHEN 0 THEN 0 ELSE stdev/mean END > 1);

SELECT inv1.w_warehouse_sk , inv1.i_item_sk , inv1.d_moy, inv1.mean,
       inv1.cov, inv2.w_warehouse_sk , inv2.i_item_sk , inv2.d_moy,
       inv2.mean, inv2.cov
  FROM inv inv1,inv inv2
 WHERE inv1.i_item_sk = inv2.i_item_sk
   AND inv1.w_warehouse_sk =  inv2.w_warehouse_sk
   AND inv1.d_moy=1
   AND inv2.d_moy=1+1
 ORDER BY inv1.w_warehouse_sk , inv1.i_item_sk ,
          inv1.d_moy, inv1.mean, inv1.cov,
          inv2.d_moy,inv2.mean, inv2.cov;

DROP VIEW IF EXISTS inv;
CREATE VIEW inv AS
```

```
    (SELECT w_warehouse_name, w_warehouse_sk, i_item_sk,
            d_moy, stdev, mean,
            CASE mean WHEN 0 THEN NULL ELSE stdev/mean END cov
       FROM (SELECT w_warehouse_name, w_warehouse_sk, i_item_sk,
                    d_moy, stddev_samp(inv_quantity_on_hand) stdev,
                    avg(inv_quantity_on_hand) mean
               FROM inventory, item, warehouse, date_dim
              WHERE inv_item_sk = i_item_sk
                AND inv_warehouse_sk = w_warehouse_sk
                AND inv_date_sk = d_date_sk
                AND d_year = 1998
              GROUP BY w_warehouse_name, w_warehouse_sk,
                    i_item_sk,d_moy) foo
      WHERE CASE mean WHEN 0 THEN 0 ELSE stdev/mean END > 1);

SELECT inv1.w_warehouse_sk, inv1.i_item_sk, inv1.d_moy,
       inv1.mean, inv1.cov, inv2.w_warehouse_sk, inv2.i_item_sk,
       inv2.d_moy, inv2.mean, inv2.cov
  FROM inv inv1, inv inv2
 WHERE inv1.i_item_sk = inv2.i_item_sk
   AND inv1.w_warehouse_sk = inv2.w_warehouse_sk
   AND inv1.d_moy= 2
   AND inv2.d_moy= 2 + 1
   AND inv1.cov > 1.5
 ORDER BY inv1.w_warehouse_sk, inv1.i_item_sk, inv1.d_moy,
          inv1.mean,inv1.cov, inv2.d_moy, inv2.mean, inv2.cov;

DROP VIEW inv;
END;
```

Listing 1.23. Query 23

Query 24. For a given product, measure the effect of competitors' prices on products' in-store and online sales. (Compute the cross-price elasticity of demand for a given product).

```
BEGIN;

CREATE VIEW competitor_price_view AS
   (SELECT i_item_sk, (imp_competitor_price - i_current_price)
           / i_current_price AS price_change, imp_start_date,
           imp_end_date - imp_start_date AS no_days
      FROM item, item_marketprices
     WHERE imp_item_sk = i_item_sk
       AND i_item_sk in (7,17)
       AND imp_competitor_price < i_current_price);

CREATE VIEW self_ws_view AS
   (SELECT ws_item_sk,
          SUM (CASE WHEN ws_sold_date_sk >= c.imp_start_date
                     AND ws_sold_date_sk < c.imp_start_date + c.no_days
                THEN ws_quantity ELSE 0 END) AS current_ws,
          SUM (CASE WHEN ws_sold_date_sk >= c.imp_start_date - c.no_days
                     AND ws_sold_date_sk < c.imp_start_date
                THEN ws_quantity ELSE 0 END) AS prev_ws
     FROM web_sales, competitor_price_view c
    WHERE ws_item_sk = c.i_item_sk
    GROUP BY 1);

CREATE VIEW self_ss_view AS
   (SELECT ss_item_sk,
          SUM (CASE WHEN ss_sold_date_sk >= c.imp_start_date
                     AND ss_sold_date_sk < c.imp_start_date + c.no_days
                THEN ss_quantity ELSE 0 END) AS current_ss,
          SUM (CASE WHEN ss_sold_date_sk >= c.imp_start_date - c.no_days
                     AND ss_sold_date_sk < c.imp_start_date
```

```
                        THEN ss_quantity ELSE 0 END) AS prev_ss
      FROM store_sales , competitor_price_view c
      WHERE c.i_item_sk = ss_item_sk
      GROUP BY 1);

SELECT i_item_sk , (current_ss + current_ws-prev_ss-prev_ws)
          / ((prev_ss + prev_ws) * price_change) AS cross_price_elasticity
  FROM competitor_price_view, self_ws_view, self_ss_view
 WHERE i_item_sk = ws_item_sk
   AND i_item_sk = ss_item_sk;

DROP VIEW self_ws_view;
DROP VIEW self_ss_view;
DROP VIEW competitor_price_view;
END;
```

Listing 1.24. Query 24

Query 25. Customer segmentation analysis: Customers are separated along the
following key shopping dimensions: recency of last visit, frequency of visits and
monetary amount. Use the store and online purchase data during a given year
to compute.

```
DROP VIEW usersegments;
CREATE VIEW usersegments AS
  (SELECT ss_customer_sk AS cid, ss_ticket_number AS oid,
          ss_sold_date_sk AS dateid, sum(ss_net_paid) AS amount
     FROM store_sales
    WHERE ss_sold_date_sk > 2452277
      AND ss_customer_sk IS NOT NULL
    GROUP BY 1,2,3
    UNION ALL
   SELECT ws_bill_customer_sk AS cid, ws_order_number AS oid,
          ws_sold_date_sk AS dateid, SUM(ws_net_paid) AS amount
     FROM web_sales
    WHERE ws_sold_date_sk > 2452277
      AND ws_bill_customer_sk is not null
      GROUP BY 1,2,3);

DROP VIEW clusteringtable;
CREATE VIEW clusteringtable AS
  (SELECT cid AS id,
          CASE WHEN 2452642 - MAX(dateid) < 60
               THEN 1.0 ELSE 0.0 END as recency,
          COUNT(oid) AS frequency,
          SUM(amount) AS totalspend
     FROM usersegments
    GROUP BY 1);

DROP TABLE user_shopping_groups;
SELECT *
  FROM kmeans (ON
        (SELECT 1)
        PARTITION BY 1
        DATABASE ('benchmark')
        USERID ('benchmark')
        PASSWORD ('benchmark')
        INPUTTABLE ('clusteringtable␣AS␣c')
        OUTPUTTABLE ('user_shopping_groups')
        NUMBERK ('8'));
```

Listing 1.25. Query 25

Query 26. Cluster customers into book buddies/ club groups based on their in store book purchasing histories.

```
CREATE VIEW clusteringtable AS
  (SELECT ss.ss_customer_sk AS cid,
          COUNT(CASE WHEN i.i_class_id=1 THEN 1 ELSE NULL END) AS id1,
          COUNT(CASE WHEN i.i_class_id=3 THEN 1 ELSE NULL END) AS id3,
          COUNT(CASE WHEN i.i_class_id=5 THEN 1 ELSE NULL END) AS id5,
          COUNT(CASE WHEN i.i_class_id=7 THEN 1 ELSE NULL END) AS id7,
          COUNT(CASE WHEN i.i_class_id=9 THEN 1 ELSE NULL END) AS id9,
          COUNT(CASE WHEN i.i_class_id=11 THEN 1 ELSE NULL END) AS id11,
          COUNT(CASE WHEN i.i_class_id=13 THEN 1 ELSE NULL END) AS id13,
          COUNT(CASE WHEN i.i_class_id=15 THEN 1 ELSE NULL END) AS id15,
          COUNT(CASE WHEN i.i_class_id=2 THEN 1 ELSE NULL END) AS id2,
          COUNT(CASE WHEN i.i_class_id=4 THEN 1 ELSE NULL END) AS id4,
          COUNT(CASE WHEN i.i_class_id=6 THEN 1 ELSE NULL END) AS id6,
          COUNT(CASE WHEN i.i_class_id=8 THEN 1 ELSE NULL END) AS id8,
          COUNT(CASE WHEN i.i_class_id=10 THEN 1 ELSE NULL END) AS id10,
          COUNT(CASE WHEN i.i_class_id=14 THEN 1 ELSE NULL END) AS id14,
          COUNT(CASE WHEN i.i_class_id=16 THEN 1 ELSE NULL END) AS id16
    FROM store_sales ss, item i
   WHERE ss.ss_item_sk = i.i_item_sk
     AND i.i_category = 'Books'
     AND ss.ss_customer_sk IS NOT NULL
   GROUP BY 1
   HAVING COUNT(ss.ss_item_sk) > 5);

SELECT *
  FROM kmeans (ON
       (SELECT 1)
       PARTITION BY 1
       DATABASE('benchmark')
       USERID('benchmark')
       PASSWORD('benchmark')
       INPUTTABLE ('clusteringtable␣AS␣c')
       OUTPUTTABLE ('book_club_groups')
       NUMBERK('2'));

SELECT clusterid, cid
  FROM kmeansplot (
       ON clusteringtable AS c
       PARTITION BY ANY
       ON book_club_groups dimension
       CENTROIDSTABLE ('book_club_groups'))
 ORDER BY clusterid, cid;

DROP TABLE IF EXISTS book_club_groups;
DROP VIEW IF EXISTS clusteringtable;
```

Listing 1.26. Query 26

Query 27. Extract competitor product names and model names (if any) from online product reviews for a given product.

```
SELECT DISTINCT *
  FROM FindNamedEntity (
    ON (SELECT pr_review_sk, pr_item_sk, pr_review_content
          FROM product_reviews
         WHERE pr_item_sk = 10653) AS p
    PARTITION BY ANY
    ON nameFind_configure AS "ConfigureTable" DIMENSION
    TEXT_COLUMN ('pr_review_content')
    MODEL('organization')
    OUTPUT_COLUMNS('pr_review_sk', 'pr_item_sk'));
```

Listing 1.27. Query 27

Query 28. Build text classifier for online review sentiment classification (positive, negative, neutral), using 60% of available reviews for training and the remaining 40% for testing. Display classifier accuracy on testing data.

```
CREATE FACT TABLE a32_trainingt (
  pr_review_sk BIGINT,
  pr_rating CHAR(3),
  pr_review_content TEXT,
  pr_item_sk BIGINT
) DISTRIBUTE BY HASH (pr_review_sk) AS
  SELECT pr_review_sk,
         (CASE pr_review_rating
           WHEN 1 THEN 'NEG'
           WHEN 2 THEN 'NEG'
           WHEN 3 THEN 'NEU'
           WHEN 4 THEN 'POS'
           WHEN 5 THEN 'POS' END) AS pr_rating,
         pr_review_content, pr_item_sk
    FROM product_reviews
  WHERE MOD (pr_review_sk, 5) IN (1,2,3);

CREATE FACT TABLE a32_testingt (
  pr_review_sk BIGINT,
  pr_rating CHAR(3),
  pr_review_content text,
  pr_item_sk BIGINT
) DISTRIBUTE BY HASH (pr_review_sk) AS
  SELECT pr_review_sk,
         (CASE pr_review_rating
           WHEN 1 THEN 'NEG'
           WHEN 2 THEN 'NEG'
           WHEN 3 THEN 'NEU'
           WHEN 4 THEN 'POS'
           WHEN 5 THEN 'POS' END) AS pr_rating,
         pr_review_content, pr_item_sk
    FROM product_reviews
  WHERE MOD (pr_review_sk, 5) IN (0, 4);

SELECT *
  FROM TextClassifierTrainer (
    ON (SELECT 1)
    PARTITION BY 1
    DATABASE('benchmark')
    USERID('benchmark')
    PASSWORD('benchmark')
    INPUTTABLE('a32_trainingt')
    TEXTCOLUMN('pr_review_content')
    CATEGORYCOLUMN('pr_rating')
    MODELFILE('senti_classifier.mod')
    CLASSIFIERTYPE('MaxEnt')
    NLPPARAMETERS('useStem:true'));

SELECT *
  FROM TextClassifier (
    ON InputTable('a32_testingt')
    TEXTCOLUMN('pr_review_content')
    MODEL('senti_classifier.mod')
    ACCUMULATE('pr_review_sk','pr_rating'));

SELECT *
  FROM TextClassifierEvaluator (
    ON TextClassifier (
        ON InputTable('a32_trainingt')
        TEXTCOLUMN('pr_review_content')
        MODEL('senti_classifier.mod')
        ACCUMULATE('pr_review_sk','pr_rating'))
    PARITION BY 1
```

```
    EXPECTCOLUMN('pr_rating')
    PREDICTCOLUMN('out_category'));

DROP TABLE a32_trainingt;
DROP TABLE a32_testingt;
```

Listing 1.28. Query 28

Query 29. Perform category affinity analysis for products purchased online together.

```
CREATE VIEW c_affinity_input AS
  (SELECT i.i_category_id AS category_cd,
          s.ws_bill_customer_sk AS customer_id
    FROM web_sales s INNER JOIN item i
      ON s.ws_item_sk = i_item_sk
    WHERE i.i_category_id IS NOT NULL);

SELECT *
  FROM cfilter (ON
        (SELECT 1)
        PARTITION BY 1
        DATABASE ('benchmark')
        USERID ('benchmark')
        PASSWORD ('benchmark')
        INPUTTABLE ('benchmark.c_affinity_input')
        OUTPUTTABLE ('c_affinity_out')
        DROPTABLE ('true')
        INPUTCOLUMNS ('category_cd')
        JOINCOLUMNS ('customer_id'));

SELECT * FROM c_affinity_out;

DROP TABLE IF EXISTS c_affinity_out;
DROP VIEW IF EXISTS c_affinity_input;
```

Listing 1.29. Query 29

Query 30. Perform category affinity analysis for products viewed together.

```
DROP VIEW IF EXISTS c_affinity_input;
CREATE VIEW c_affinity_input AS
  (SELECT i.i_category_id AS category_cd,
          s.wcs_user_sk AS customer_id
    FROM web_clickstreams s INNER JOIN item i
      ON s.wcs_item_sk = i_item_sk
    WHERE s.wcs_item_sk IS NOT NULL
      AND i.i_category_id IS NOT NULL
      AND s.wcs_user_sk IS NOT NULL);

SELECT *
  FROM cfilter (ON
        (SELECT 1)
        PARTITION BY 1
        DATABASE ('benchmark')
        USERID ('benchmark')
        PASSWORD ('benchmark')
        INPUTTABLE ('benchmark.c_affinity_input')
        OUTPUTTABLE ('c_affinity_out')
        DROPTABLE ('true')
        INPUTCOLUMNS ('category_cd')
        JOINCOLUMNS ('customer_id'));

SELECT *
```

```
  FROM c_affinity_out;

DROP VIEW IF EXISTS c_affinity_input;
DROP TABLE IF EXISTS c_affinity_out;
```

Listing 1.30. Query 30

B BigBench Schema

Below is the complete schema definition for BigBench in Teradata Aster DBMS
syntax.

```
DROP TABLE IF EXISTS customer_simple;
DROP TABLE IF EXISTS customer_addr_simple;
DROP TABLE IF EXISTS inventory_simple;
DROP TABLE IF EXISTS item_simple;
DROP TABLE IF EXISTS store_sales_simple;
DROP TABLE IF EXISTS store_returns_simple;
DROP TABLE IF EXISTS web_sales_simple;
DROP TABLE IF EXISTS web_returns_simple;


DROP TABLE IF EXISTS customer cascade;
DROP TABLE IF EXISTS customer_address cascade;
DROP TABLE IF EXISTS customer_demographics cascade;
DROP TABLE IF EXISTS date_dim cascade;
DROP TABLE IF EXISTS dbgen_version cascade;
DROP TABLE IF EXISTS household_demographics cascade;
DROP TABLE IF EXISTS income_band cascade;
DROP TABLE IF EXISTS item cascade;
DROP TABLE IF EXISTS promotion cascade;
DROP TABLE IF EXISTS reason cascade;
DROP TABLE IF EXISTS ship_mode cascade;
DROP TABLE IF EXISTS store cascade;
DROP TABLE IF EXISTS time_dim cascade;
DROP TABLE IF EXISTS warehouse cascade;
DROP TABLE IF EXISTS web_site cascade;
DROP TABLE IF EXISTS web_page cascade;
DROP TABLE IF EXISTS inventory cascade;
DROP TABLE IF EXISTS store_sales cascade;
DROP TABLE IF EXISTS store_returns cascade;
DROP TABLE IF EXISTS web_sales cascade;
DROP TABLE IF EXISTS web_returns cascade;

CREATE TABLE dbgen_version (
   dv_version        VARCHAR(16),
   dv_create_date    date,
   dv_create_time    time,
   dv_cmdline_args   VARCHAR(200)
   ) DISTRIBUTE BY REPLICATION;

CREATE TABLE customer_demographics (
   cd_demo_sk            BIGINT NOT NULL,
   cd_gender             CHAR(1),
   cd_marital_status     CHAR(1),
   cd_education_status   CHAR(20),
   cd_purchase_estimate  INTEGER,
   cd_credit_rating      CHAR(10),
   cd_dep_count          INTEGER,
   cd_dep_employed_count INTEGER,
   cd_dep_college_count  INTEGER,
   PRIMARY KEY (cd_demo_sk)
   ) DISTRIBUTE BY REPLICATION;
```

```
CREATE TABLE date_dim (
    d_date_sk              BIGINT   NOT NULL,
    d_date_id              CHAR(16) NOT NULL,
    d_date                 DATE,
    d_month_seq            INTEGER,
    d_week_seq             INTEGER,
    d_quarter_seq          INTEGER,
    d_year                 INTEGER,
    d_dow                  INTEGER,
    d_moy                  INTEGER,
    d_dom                  INTEGER,
    d_qoy                  INTEGER,
    d_fy_year              INTEGER,
    d_fy_quarter_seq       INTEGER,
    d_fy_week_seq          INTEGER,
    d_day_name             CHAR(9),
    d_quarter_name         CHAR(6),
    d_holiday              CHAR(1),
    d_weekend              CHAR(1),
    d_following_holiday    CHAR(1),
    d_first_dom            INTEGER,
    d_last_dom             INTEGER,
    d_same_day_ly          INTEGER,
    d_same_day_lq          INTEGER,
    d_current_day          CHAR(1),
    d_current_week         CHAR(1),
    d_current_month        CHAR(1),
    d_current_quarter      CHAR(1),
    d_current_year         CHAR(1),
    PRIMARY KEY (d_date_sk)
) DISTRIBUTE BY REPLICATION;

CREATE TABLE warehouse (
    w_warehouse_sk    BIGINT   NOT NULL,
    w_warehouse_id    CHAR(16) NOT NULL,
    w_warehouse_name  VARCHAR(20),
    w_warehouse_sq_ft INTEGER,
    w_street_number   CHAR(10),
    w_street_name     VARCHAR(60),
    w_street_type     CHAR(15),
    w_suite_number    CHAR(10),
    w_city            VARCHAR(60),
    w_county          VARCHAR(30),
    w_state           CHAR(2),
    w_zip             CHAR(10),
    w_country         VARCHAR(20),
    w_gmt_offset      DECIMAL(5,2),
    PRIMARY KEY (w_warehouse_sk)
) DISTRIBUTE BY REPLICATION;

CREATE TABLE ship_mode (
    sm_ship_mode_sk BIGINT   NOT NULL,
    sm_ship_mode_id CHAR(16) NOT NULL,
    sm_type         CHAR(30),
    sm_code         CHAR(10),
    sm_carrier      CHAR(20),
    sm_contract     CHAR(20),
    PRIMARY KEY (sm_ship_mode_sk)
) DISTRIBUTE BY REPLICATION;

CREATE TABLE time_dim (
    t_time_sk   BIGINT   NOT NULL,
    t_time_id   CHAR(16) NOT NULL,
    t_time      INTEGER,
    t_hour      INTEGER,
    t_minute    INTEGER,
    t_second    INTEGER,
    t_am_pm     CHAR(2),
```

```
 t_shift      CHAR(20),
 t_sub_shift CHAR(20),
 t_meal_time CHAR(20),
 PRIMARY KEY (t_time_sk)
 ) DISTRIBUTE BY REPLICATION;

CREATE TABLE reason (
 r_reason_sk  BIGINT   NOT NULL,
 r_reason_id  CHAR(16) NOT NULL,
 r_reason_desc CHAR(100),
 PRIMARY KEY (r_reason_sk)
 ) DISTRIBUTE BY REPLICATION;

CREATE TABLE income_band (
 ib_income_band_sk BIGINT NOT NULL,
 ib_lower_bound    INTEGER,
 ib_upper_bound    INTEGER,
 PRIMARY KEY (ib_income_band_sk)
 ) DISTRIBUTE BY REPLICATION;

CREATE TABLE store (
 s_store_sk          BIGINT   NOT NULL,
 s_store_id          CHAR(16) NOT NULL,
 s_rec_start_date    DATE,
 s_rec_end_date      DATE,
 s_closed_date_sk    BIGINT,
 s_store_name        VARCHAR(50),
 s_number_employees  INTEGER,
 s_floor_space       INTEGER,
 s_hours             CHAR(20),
 s_manager           VARCHAR(40),
 s_market_id         INTEGER,
 s_geography_class   VARCHAR(100),
 s_market_desc       VARCHAR(100),
 s_market_manager    VARCHAR(40),
 s_division_id       INTEGER,
 s_division_name     VARCHAR(50),
 s_company_id        INTEGER,
 s_company_name      VARCHAR(50),
 s_street_number     VARCHAR(10),
 s_street_name       VARCHAR(60),
 s_street_type       CHAR(15)
 s_suite_number      CHAR(10),
 s_city              VARCHAR(60),
 s_county            VARCHAR(30),
 s_state             CHAR(2),
 s_zip               CHAR(10),
 s_country           VARCHAR(20),
 s_gmt_offset        DECIMAL(5,2),
 s_tax_precentage    DECIMAL(5,2),
 PRIMARY KEY (s_store_sk)
 ) DISTRIBUTE BY REPLICATION;

CREATE TABLE web_site (
 web_site_sk             BIGINT   NOT NULL,
 web_site_id             CHAR(16) NOT NULL,
 web_rec_start_date      DATE,
 web_rec_end_date        DATE,
 web_name                VARCHAR(50),
 web_open_date_sk        BIGINT,
 web_close_date_sk       BIGINT,
 web_class               VARCHAR(50),
 web_manager             VARCHAR(40),
 web_mkt_id              INTEGER,
 web_mkt_class           VARCHAR(50),
 web_mkt_desc            VARCHAR(100),
 web_market_manager      VARCHAR(40),
 web_company_id          INTEGER,
```

```
web_company_name          CHAR(50),
web_street_number         CHAR(10),
web_street_name           VARCHAR(60),
web_street_type           CHAR(15),
web_suite_number          CHAR(10),
web_city                  VARCHAR(60),
web_county                VARCHAR(30),
web_state                 CHAR(2),
web_zip                   CHAR(10),
web_country               VARCHAR(20),
web_gmt_offset            DECIMAL(5,2),
web_tax_percentage        DECIMAL(5,2),
PRIMARY KEY (web_site_sk)
) DISTRIBUTE BY REPLICATION;

CREATE TABLE household_demographics (
hd_demo_sk                BIGINT NOT NULL,
hd_income_band_sk         BIGINT,
hd_buy_potential          CHAR(15),
hd_dep_count              INTEGER,
hd_vehicle_count          INTEGER,
PRIMARY KEY (hd_demo_sk)
) DISTRIBUTE BY REPLICATION;

CREATE TABLE web_page (
wp_web_page_sk            BIGINT   NOT NULL,
wp_web_page_id            CHAR(16) NOT NULL,
wp_rec_start_date         DATE,
wp_rec_end_date           DATE,
wp_creation_date_sk       BIGINT,
wp_access_date_sk         BIGINT,
wp_autogen_flag           CHAR(1),
wp_customer_sk            BIGINT,
wp_url                    VARCHAR(100),
wp_type                   CHAR(50),
wp_char_count             INTEGER,
wp_link_count             INTEGER,
wp_image_count            INTEGER,
wp_max_ad_count           INTEGER,
PRIMARY KEY (wp_web_page_sk)
) DISTRIBUTE BY REPLICATION;

CREATE TABLE promotion (
p_promo_sk                BIGINT   NOT NULL,
p_promo_id                CHAR(16) NOT NULL,
p_start_date_sk           BIGINT,
p_end_date_sk             BIGINT,
p_item_sk                 BIGINT,
p_cost                    DECIMAL(15,2),
p_response_target         INTEGER,
p_promo_name              CHAR(50),
p_channel_dmail           CHAR(1),
p_channel_email           CHAR(1),
p_channel_catalog         CHAR(1),
p_channel_tv              CHAR(1),
p_channel_radio           CHAR(1),
p_channel_press           CHAR(1),
p_channel_event           CHAR(1),
p_channel_demo            CHAR(1),
p_channel_details         VARCHAR(100),
p_purpose                 CHAR(15),
p_discount_active         CHAR(1),
PRIMARY KEY (p_promo_sk)
) DISTRIBUTE BY REPLICATION;

CREATE TABLE customer (
c_customer_sk             BIGINT   NOT NULL,
c_customer_id             CHAR(16) NOT NULL,
```

```
    c_current_cdemo_sk          BIGINT,
    c_current_hdemo_sk          BIGINT,
    c_current_addr_sk           BIGINT,
    c_first_shipto_date_sk      BIGINT,
    c_first_sales_date_sk       BIGINT,
    c_salutation                CHAR(10),
    c_first_name                CHAR(20),
    c_last_name                 CHAR(30),
    c_preferred_cust_flag       CHAR(1),
    c_birth_day                 INTEGER,
    c_birth_month               INTEGER,
    c_birth_year                INTEGER,
    c_birth_country             VARCHAR(20),
    c_login                     CHAR(13),
    c_email_address             CHAR(50),
    c_last_review_date          CHAR(10),
    PRIMARY KEY (c_customer_sk)
) DISTRIBUTE BY HASH (c_customer_sk);

CREATE TABLE customer_address (
    ca_address_sk               BIGINT    NOT NULL,
    ca_address_id               CHAR(16)  NOT NULL,
    ca_street_number            CHAR(10),
    ca_street_name              VARCHAR(60),
    ca_street_type              CHAR(15),
    ca_suite_number             CHAR(10),
    ca_city                     VARCHAR(60),
    ca_county                   VARCHAR(30),
    ca_state                    CHAR(2),
    ca_zip                      CHAR(10),
    ca_country                  VARCHAR(20),
    ca_gmt_offset               DECIMAL(5,2),
    ca_location_type            CHAR(20),
    PRIMARY KEY (ca_address_sk)
) DISTRIBUTE BY HASH (ca_address_sk);

CREATE TABLE inventory (
    inv_date_sk                 BIGINT NOT NULL,
    inv_item_sk                 BIGINT NOT NULL,
    inv_warehouse_sk            BIGINT NOT NULL,
    inv_quantity_on_hand        INTEGER
) DISTRIBUTE BY HASH (inv_item_sk);

CREATE TABLE item (
    i_item_sk                   BIGINT    NOT NULL,
    i_item_id                   CHAR(16)  NOT NULL,
    i_rec_start_date            DATE,
    i_rec_end_date              DATE,
    i_item_desc                 VARCHAR(200),
    i_current_price             DECIMAL(7,2),
    i_wholesale_cost            DECIMAL(7,2),
    i_brand_id                  INTEGER,
    i_brand                     CHAR(50),
    i_class_id                  INTEGER,
    i_class                     CHAR(50),
    i_category_id               INTEGER,
    i_category                  CHAR(50),
    i_manufact_id               INTEGER,
    i_manufact                  CHAR(50),
    i_size                      CHAR(20),
    i_formulation               CHAR(20),
    i_color                     CHAR(20),
    i_units                     CHAR(10),
    i_container                 CHAR(10),
    i_manager_id                INTEGER,
    i_product_name              CHAR(50),
    PRIMARY KEY (i_item_sk)
) DISTRIBUTE BY HASH (i_item_sk);
```

```
CREATE TABLE store_sales (
  ss_sold_date_sk          BIGINT default 9999999,
  ss_sold_time_sk          BIGINT,
  ss_item_sk               BIGINT NOT NULL,
  ss_customer_sk           BIGINT,
  ss_cdemo_sk              BIGINT,
  ss_hdemo_sk              BIGINT,
  ss_addr_sk               BIGINT,
  ss_store_sk              BIGINT,
  ss_promo_sk              BIGINT,
  ss_ticket_number         BIGINT NOT NULL,
  ss_quantity              INTEGER,
  ss_wholesale_cost        DECIMAL(7,2),
  ss_list_price            DECIMAL(7,2),
  ss_sales_price           DECIMAL(7,2),
  ss_ext_discount_amt      DECIMAL(7,2),
  ss_ext_sales_price       DECIMAL(7,2),
  ss_ext_wholesale_cost    DECIMAL(7,2),
  ss_ext_list_price        DECIMAL(7,2),
  ss_ext_tax               DECIMAL(7,2),
  ss_coupon_amt            DECIMAL(7,2),
  ss_net_paid              DECIMAL(7,2),
  ss_net_paid_inc_tax      DECIMAL(7,2),
  ss_net_profit            DECIMAL(7,2),
) DISTRIBUTE BY HASH (ss_item_sk);

CREATE TABLE store_returns (
  sr_returned_date_sk      BIGINT default 9999999,
  sr_return_time_sk        BIGINT,
  sr_item_sk               BIGINT NOT NULL,
  sr_customer_sk           BIGINT,
  sr_cdemo_sk              BIGINT,
  sr_hdemo_sk              BIGINT,
  sr_addr_sk               BIGINT,
  sr_store_sk              BIGINT,
  sr_reason_sk             BIGINT,
  sr_ticket_number         BIGINT NOT NULL,
  sr_return_quantity       INTEGER,
  sr_return_amt            DECIMAL(7,2),
  sr_return_tax            DECIMAL(7,2),
  sr_return_amt_inc_tax    DECIMAL(7,2),
  sr_fee                   DECIMAL(7,2),
  sr_return_ship_cost      DECIMAL(7,2),
  sr_refunded_cash         DECIMAL(7,2),
  sr_reversed_charge       DECIMAL(7,2),
  sr_store_credit          DECIMAL(7,2),
  sr_net_loss              DECIMAL(7,2),
) DISTRIBUTE BY HASH (sr_item_sk);

CREATE TABLE web_sales (
  ws_sk                    BIGINT NOT NULL,
  ws_sold_date_sk          BIGINT default 9999999,
  ws_sold_time_sk          BIGINT,
  ws_ship_date_sk          BIGINT,
  ws_item_sk               BIGINT NOT NULL,
  ws_bill_customer_sk      BIGINT,
  ws_bill_cdemo_sk         BIGINT,
  ws_bill_hdemo_sk         BIGINT,
  ws_bill_addr_sk          BIGINT,
  ws_ship_customer_sk      BIGINT,
  ws_ship_cdemo_sk         BIGINT,
  ws_ship_hdemo_sk         BIGINT,
  ws_ship_addr_sk          BIGINT,
  ws_web_page_sk           BIGINT,
  ws_web_site_sk           BIGINT,
  ws_ship_mode_sk          BIGINT,
  ws_warehouse_sk          BIGINT,
```

```
    ws_promo_sk                 BIGINT,
    ws_order_number             BIGINT NOT NULL,
    ws_quantity                 INTEGER,
    ws_wholesale_cost           DECIMAL(7,2),
    ws_list_price               DECIMAL(7,2),
    ws_sales_price              DECIMAL(7,2),
    ws_ext_discount_amt         DECIMAL(7,2),
    ws_ext_sales_price          DECIMAL(7,2),
    ws_ext_wholesale_cost       DECIMAL(7,2),
    ws_ext_list_price           DECIMAL(7,2),
    ws_ext_tax                  DECIMAL(7,2),
    ws_coupon_amt               DECIMAL(7,2),
    ws_ext_ship_cost            DECIMAL(7,2),
    ws_net_paid                 DECIMAL(7,2),
    ws_net_paid_inc_tax         DECIMAL(7,2),
    ws_net_paid_inc_ship        DECIMAL(7,2),
    ws_net_paid_inc_ship_tax    DECIMAL(7,2),
    ws_net_profit               DECIMAL(7,2),
    ) DISTRIBUTE BY HASH (ws_sk);

CREATE TABLE web_returns (
    wr_returned_date_sk         BIGINT default 9999999,
    wr_returned_time_sk         BIGINT,
    wr_item_sk                  BIGINT NOT NULL,
    wr_refunded_customer_sk     BIGINT,
    wr_refunded_cdemo_sk        BIGINT,
    wr_refunded_hdemo_sk        BIGINT,
    wr_refunded_addr_sk         BIGINT,
    wr_returning_customer_sk    BIGINT,
    wr_returning_cdemo_sk       BIGINT,
    wr_returning_hdemo_sk       BIGINT,
    wr_returning_addr_sk        BIGINT,
    wr_web_page_sk              BIGINT,
    wr_reason_sk                BIGINT,
    wr_order_number             BIGINT NOT NULL,
    wr_return_quantity          INTEGER,
    wr_return_amt               DECIMAL(7,2),
    wr_return_tax               DECIMAL(7,2),
    wr_return_amt_inc_tax       DECIMAL(7,2),
    wr_fee                      DECIMAL(7,2),
    wr_return_ship_cost         DECIMAL(7,2),
    wr_refunded_cash            DECIMAL(7,2),
    wr_reversed_charge          DECIMAL(7,2),
    wr_account_credit           DECIMAL(7,2),
    wr_net_loss                 DECIMAL(7,2),
    ) DISTRIBUTE BY HASH (wr_item_sk);

DROP TABLE IF EXISTS item_marketprices cascade;
DROP TABLE IF EXISTS web_clickstreams cascade;
DROP TABLE IF EXISTS product_reviews cascade;

CREATE TABLE item_marketprices (
    imp_sk                      BIGINT NOT NULL,
    imp_item_sk                 BIGINT NOT NULL,
    imp_competitor              VARCHAR(20),
    imp_competitor_price        DECIMAL(7,2),
    imp_start_date              BIGINT,
    imp_end_date                BIGINT,
    PRIMARY KEY (imp_sk)
    ) DISTRIBUTE BY HASH (imp_sk);

CREATE TABLE web_clickstreams (
    wcs_click_sk                BIGINT NOT NULL,
    wcs_click_date_sk           BIGINT,
    wcs_click_time_sk           BIGINT,
    wcs_sales_sk                BIGINT,
    wcs_item_sk                 BIGINT,
    wcs_web_page_sk             BIGINT,
```

```
    wcs_user_sk               BIGINT,
    PRIMARY KEY (wcs_click_sk)
    ) DISTRIBUTE BY HASH (wcs_click_sk);

CREATE TABLE product_reviews (
    pr_review_sk              BIGINT NOT NULL,
    pr_review_date            DATE,
    pr_review_time            CHAR(6),
    pr_review_rating          INT    NOT NULL,
    pr_item_sk                BIGINT NOT NULL,
    pr_user_sk                BIGINT,
    pr_order_sk               BIGINT,
    pr_review_content         TEXT    NOT NULL,
    PRIMARY KEY (pr_review_sk)
    ) DISTRIBUTE BY HASH (pr_review_sk);
```

Author Index